LINEAR ALGEBRA WITH DERIVE®

BENNY EVANS
OKLAHOMA STATE UNIVERSITY

JERRY JOHNSON
OKLAHOMA STATE UNIVERSITY

JOHN WILEY & SONS, INC.
New York Chichester Brisbane Toronto Singapore

ISBN 0-471-59194-7

Printed in the United States of America

10 9 8 7 6 5 4 3 2 1

Table of Contents

Preface v

1. SYSTEMS OF EQUATIONS 1

   • Solutions of systems of equations

2. AUGMENTED MATRICES AND ELEMENTARY ROW OPERATIONS 17

   • Augmented matrix • Elementary row operation • Echelon form • Reduced echelon form • Gauss-Jordan Elimination • Transpose • Rank

3. THE ALGEBRA OF MATRICES 37

   • Matrix addition • Scalar multiplication • Matrix multiplication

4. INVERSES OF MATRICES 53

   • Matrix inversion

5. DETERMINANTS, ADJOINTS, AND CRAMER'S RULE 63

   • Determinant • Adjoint • Cramer's rule

6. APPLICATION: MATRIX ALGEBRA AND MODULAR ARITHMETIC 75

   • Modular arithmetic • Matrix operations • Hill codes

7. VECTOR PRODUCTS, LINES, AND PLANES 87

   • Dot product • Cross product • Projection • Unit vector • Vectors in R^n • Orthogonal vectors

8. VECTOR SPACES AND SUBSPACES 103

   • Vector space • Subspace • Spaces of functions and matrices • Linear combination • Spanning set • Null space • Rank of a matrix

9. INDEPENDENCE, BASIS AND, DIMENSION 119

   • Linearly independent set • Basis • Dimension • Coordinate vector

10. ROW SPACE, COLUMN SPACE, AND NULL SPACE 133

    • Row space • Column space • Null space • Rank • Nullity

11. INNER PRODUCT SPACES 143

    • General inner products

PREFACE

This is an enrichment supplement to a traditional introductory linear algebra course. Its purpose is to help students (and teachers) use the *DERIVE*® program as a tool to solve problems that arise in such a course. This book adds a new dimension to the conventional linear algebra class by providing problems that go beyond the level of rote calculations and template exercises, but at the same time fit comfortably into the traditional course.

Here's an example: Describe all matrices that commute with a given 3 by 3 matrix. (See Exercise 7 in Chapter 3.) This exercise makes the student *use* the concept of commutativity which leads to a system of nine equations in nine unknowns, a daunting task to solve with nothing but pencil and paper, but *DERIVE* will do it almost instantly. The student must then use the solution set in order to describe the required set of matrices. This exercise is typical of many in this text in that students are not asked to simply do a calculation with *DERIVE*, they must actually solve a problem.

Key Features

- No prior knowledge of *DERIVE* is required. Appendix I summarizes important commands and will help novice users get started. Students with a little *DERIVE* experience should be able to begin in Chapter I.
- Most problems go beyond the level of rote calculations and "template" exercises. The exercise sets start with examples that are similar to the solved problems and progress to more challenging problems.
- Numerous applications are provided including optimization, Markov chains, systems of differential equations, the Gauss-Seidel method, generalized inverses, curve fitting, rotation of axes, and cryptography.
- The *LU* and *QR* factorizations are presented.
- Many of the recommendations of the Linear Algebra Curriculum Study Group are followed.
- Optional code for automating some procedures is provided in Appendix II.
- Many of the problems have been class tested by the authors and others.

For information about *DERIVE* contact CipherSystems, 3468 San Juan Circle, Reno, NV 89509. (702) 329-4424.

Structure of the Book

This manual is written to be used with any linear algebra text, but we were guided by the contents, terminology and general flow of Howard Anton's *Elementary Linear Algebra.* Each chapter has five parts: Introduction, Solved Problems, Exercises, Exploration and Discovery, and Laboratory Exercises. **Solved Problems** are examples that provide a context for *DERIVE* instructions that one is likely to encounter in the chapter and may also contain useful mathematics hints as well. **Exercises** are problems that generally ask for well defined answers. They include unusual problems, such as the one mentioned in the second paragraph, as well as variations on conventional exercises that are sufficiently complex that solving them without assistance from a computer is not practical. **Exploration and Discovery** problems are those that go a bit beyond the conventional exercise. Some are more involved and difficult. Some may not simply ask for an answer, but invite one to explore a situation and then to discuss observations and findings. **Laboratory Exercises** are problems with structured responses suitable for use as a lab assignment. They can be detached and handed in.

Most *DERIVE* instructions are provided in the context of problems to be solved, but additional help is provided in Appendix I. Appendix II contains optional code for *DERIVE* files that students or instructors may find useful.

This should not be considered a substitute for the *DERIVE User Manual.* We have tried to include enough *DERIVE* instructions and suggestions in both the solved problems and Appendix I so that the reader will not have to spend a lot of time referring to the user manual, but ultimately one should consult it for details.

The *DERIVE*® Program

> *The novice who has never used DERIVE before should begin by reading* Getting Started *in Appendix I.*

DERIVE is a comprehensive computer algebra system that requires only a 512K PC-compatible. As of the printing of this book, there is no Macintosh version, but *DERIVE* will run under MicroSoft Windows as a full-screen DOS application. There is also a version that runs on HP "palm-top" computers.

Any sophisticated software takes some practice and experience to master, but we have used *DERIVE* in our courses almost since its first release and are convinced that one of its strongest educational advantages is that it is easy for students to learn and to use. It is also powerful enough for professional applications in mathematics, science, and engineering.

This manual was written using *DERIVE* version 2.5. If you are using an earlier version, you will notice a few differences. We have tried to mention them in the rare occasions when they might cause confusion. In particular, we will use the new APPEND function, but we have provided code in Appendix II for earlier versions. The most conspicuous new feature in version 2.5 is in the **Plot** command. See Sections 13 and 14 in Appendix I.

After this manuscript was composed, version 2.55 appeared in which the arrow keys may be used to move the cursor on the **Author** line. See Section 8 in Appendix I.

The most important thing to be alert to is this: *DERIVE* distinguishes between the vector [2, 3] and the 1 by 2 matrix [2 3]. In *DERIVE*, a matrix is a vector whose entries are other vectors. If you **Author** `[2,3]`, *DERIVE* accepts it as a vector. If you **Author** `[[2,3]]`, *DERIVE* accepts it as a matrix. For example, *DERIVE* will take the transpose of a matrix, but not a vector! (The syntax for the transpose of A is A‘.) Thus, *DERIVE* will compute A times [2, 3] or A times [[2, 3]]‘ correctly but *not* A times [2, 3]‘. For examples and further discussion, see Section 4 in Appendix I.

Suggestions for Incorporating *DERIVE*: What Has Worked for Us

- We normally spend a period with our classes at the computer laboratory in the first week or two of the semester to introduce them to *DERIVE*. (Our experience is that students quickly become comfortable with *DERIVE* and that only minimal help is needed later.)

- After this initial session we may return to the lab with the class two or three more times during the semester. It is important to try to integrate the lab experience with material that is currently being covered in the course. We often begin such a lab session by asking students to work through a solved problem and later give them a quiz that consists of a similar exercise. Grades in the lab are important. Even the best students appreciate credit for their work, and there are always students who will remain passive in the lab without the incentive of a grade.

- We give outside lab assignments from time to time. The frequency may vary from five to ten assignments a semester. We encourage students to work together on the assignments, but we insist that they turn in their own printouts and write up the results in their own words.

- We have found that some classroom discussion of an assigned exercise is desirable, including a few words about new *DERIVE* commands. However, all a student needs should be in the **Solved Problems** and Appendix I.

- To avoid extra expense to the student, a school's lab might buy enough manuals for one class (say 30 or 40) and check them out to students during lab time.

Important Conventions

You will notice early in this book the issue of uppercase versus lowercase letters. We encourage you to read Section 2 of Appendix I.

We have tried to follow three notational conventions in this manual.

1. When a key is to be depressed on the keyboard, it will be put in a box. For example, [F3] [Enter] means you are to press the function key F3 and then press the Enter key in sequence. Sometimes you may have to hold down two keys at once. In this case, both keys will appear in same box with a space between them. For example, to type the Greek letter π, you hold down the key marked Alt while you press the letter P. We will denote this by [Alt P]. Similarly, [Ctrl →] means you hold down the key marked Ctrl while you press the right arrow key.

2. *DERIVE* is menu driven. Commands are issued by selecting a key word at the bottom of the screen and typing the capital letter that appears in it, or by highlighting it with the space bar and pressing [Enter]. Such commands appear in boldface and are spelled as they appear in the *DERIVE* menu. For example, if you see **Simplify**, you would press the letter [S] to give *DERIVE* the command to simplify an expression. To approximate an expression, the command is **approX** and you would press the [X] key.

   Sometimes a sequence of commands is necessary, in which case we may list them. For example, to set *DERIVE* so that it will report answers in decimal form, we would say "use the commands **Options Notation Decimal**."

3. When you must enter an expression into *DERIVE*, it will be put in what is sometimes called "typewriter" or "teletype" style. In this case, you type exactly the symbols you see except the period at the end. For example, if we want you to enter $\frac{x^2}{3}$, we will say **Author** `x^2/3`. (**Author** is the command that prepares *DERIVE* to accept input.) To have you enter the vector (2, 3) we may say **Author** `[2,3]`.

A Word to Students

DERIVE has essentially automated most of the standard algebraic and matrix calculations you will encounter, just as scientific calculators have done with arithmetic. It will simplify complicated expressions, solve equations, and draw graphs. It will also row-reduce matrices and find their determinants and inverses. But that doesn't mean linear algebra is obsolete or unimportant. Even though calculators will do arithmetic, we still have to know what questions to ask, understand what the answers mean, and realize when an obvious error has been made. In the same way, we still have to understand the definitions, concepts, and processes that are involved in linear algebra so we will know what to tell *DERIVE* to do, what its answers mean, and be able to detect errors. *DERIVE* only does the calculations; you must still do the thinking.

Always view any computer or calculator output critically. Be alert for answers that seem strange; you might have hit the wrong key, entered the wrong data, or made some other mistake. It is even possible that the program has a bug! If a problem asks for the cost of materials to make a shoe box and you get $123.28 or [4, 9] you should suspect something is wrong!

Clear communication is at least as important in mathematics as in other fields. You should always write your answers neatly in complete, logical sentences. Re-read what you have written and ask yourself if it really makes sense.

Before you turn on the computer, work through as much of an assigned problem as you can with pencil and paper, taking note of exactly where you think the computer will be required and for what purpose. You may be surprised at how little time you will actually have to spend in front of the machine if you follow this advice.

With few exceptions, the problems in the sections entitled **Exercises** are traditional in that they ask for a clearly defined answer; however, in the **Exploration and Discovery** sections this may not be the case. Here we may not simply ask for an answer, but invite you to explore a situation and then to discuss your observations and findings. The instructions may even be vague on occasion. That's what discovery is all about. In fact, if your interpretation and analysis are correct, but not what we or your teacher expected, there is no cause for concern. On the contrary, it is a sign that you are indeed discovering mathematics.

A Word to Instructors

Each chapter has a section entitled **Solved Problems** that may contain a straightforward example to provide a "warm-up problem" or a context for illustrating *DERIVE* commands and hints that may be useful later. However, we have largely avoided exercises that just show off *DERIVE's* power with no apparent mathematical lesson (find the inverse of the 10 by 10 Hilbert matrix $\{\frac{1}{i+j-1}\}_{i,j=1}^{10}$) or routine exercises that can be done easily by hand (row-reduce $\begin{bmatrix} 1 & 2 \\ -2 & -4 \end{bmatrix}$). Our opinion is that the computer should be a tool, not a crutch. To ask students to appeal to *DERIVE* for simple problems sends the wrong message, much like suggesting they use a calculator to divide 4 by 2.

Some of the problems in the **Exploration and Discovery** sections do not simply ask for an answer, but invite students to explore and then discuss their observations. In some cases we have deliberately avoided telling the students exactly what is expected of them. You may supply hints, guidance, or suggestions as you see fit.

We have only used real numbers in the exercises in this book, but *DERIVE* works in the complex field. It is quite easy to include complex numbers in some examples if you so desire. Read Section 16 in Appendix I for information on handling complex numbers with *DERIVE.*

Finally, there is some *DERIVE* code we have included in Appendix II that does specialized calculations, such as *LU* and *QR* factorizations. They might be too tedious for most students to enter, but you may want to do it yourself, save them for future use, and supply them to your students.

ACKNOWLEDGMENT

We would like to express our appreciation to Al Rich and David Stoutemyer, co-authors of the *DERIVE*® program, for reviewing the original manuscript. Their suggestions and corrections were very helpful in improving the final version.

We would also like to thank Professor Kent Nagle of the University of South Florida for class testing this material using Maple® and for providing corrections and student reactions.

CHAPTER 1

SYSTEMS OF EQUATIONS

LINEAR ALGEBRA CONCEPTS

- **Solutions of systems of equations**

Introduction

Elementary linear algebra commonly begins with a study of systems of linear equations. The arithmetic required to solve such systems is often quite formidable, especially if the system involves many variables and equations or if the coefficients are complicated. Sometimes there is a unique solution, but often there may be no solution at all or a whole *set* of solutions whose description must be determined. *DERIVE* can handle each of these three cases, allowing attention to be focused on the utility of systems of equations rather than on the tedious calculations involved in solving them.

Solved Problems

Solved Problem 1: Solve the following system of equations.

$$\begin{aligned} 3x + 4y - 7z &= 8 \\ 2x - 3y + 4z &= 2 \\ 4x + 2y - 3z &= 4 \end{aligned}$$

Solution: To solve a system of linear equations with *DERIVE*, enter the equations in a list separated by commas and enclosed by square brackets. In our example we **Author** `[3x+4y-7z=8,2x-3y+4z=2,4x+2y-3z=4]`, ask *DERIVE* to **soLve**, and we see the unique solution in expression 2 of Figure 1.1.

```
1:    [3 x + 4 y - 7 z = 8, 2 x - 3 y + 4 z = 2, 4 x + 2 y - 3 z = 4]

        ┌     16          110          80 ┐
2:      │x = ───, y = - ───, z = - ─── │
        └     21          21           21 ┘
```

Figure 1.1: **A system of equations with a unique solution**

DERIVE hint: An alternative method for entering the system of equations is to **Author** each equation so that it appears as a separate expression (say as #7, #8, and #9) and then **Author** `[#7,#8,#9]`.

Solved Problem 2: Solve the following system of equations. The system has an infinite number of solutions; find a specific solution with $x > 10$.

$$\begin{aligned} 3x + 4y + z &= 8 \\ 2x - y + 3z &= 2 \end{aligned}$$

Solution: **Author** `[3x+4y+z=8,2x-y+3z=2]`. *DERIVE* recognizes that there are more variables than equations, so when we **soLve** we are asked which variables we want to solve for. (*DERIVE* refers to these as the *solve variables*). If we accept the defaults (x and y) by pressing Enter twice, *DERIVE* will present the solution in expression 2 of Figure 1.2. This shows that there are infinitely many solutions – one for each value of z we choose.

When a system of equations has infinitely many solutions, they are expressed in terms of one or more variables called *parameters*. In this case, z itself can be used as the parameter, or we can introduce another symbol. Below we have expressed the solution in terms of s.

$$\begin{aligned} x &= \frac{16 - 13s}{11} \\ y &= \frac{7s + 10}{11} \\ z &= s \end{aligned}$$

Solutions are presented in this way to emphasize the fact that the parameter s can be assigned any value at all, and each assignment determines values for x, y, and z that comprise a solution of the system.

```
1:   [3 x + 4 y + z = 8, 2 x - y + 3 z = 2]

        ┌      16 - 13 z         7 z + 10 ┐
2:      │x = ───────────, y = ──────────│
        └        11                11     ┘

        ┌      16 - 13 (-8)         7 (-8) + 10 ┐
3:      │x = ─────────────, y = ─────────────│
        └         11                   11       ┘

        ┌     120           46 ┐
4:      │x = ─────, y = - ────│
        └     11            11 ┘
```

Figure 1.2: **A system of equations with many solutions**

To determine a specific solution of the system with $x > 10$, we want to choose a value of s that makes $\dfrac{16 - 13s}{11} > 10$. The solution of this inequality is $s < -\frac{94}{13}$. (*DERIVE* will do it, but it's easy to solve by hand.) Thus, *any* value of s less than $-\frac{94}{13}$ will do; we choose $s = -8$.

To find our final answer, we must substitute -8 for z. Highlight expression 2 and use **Manage Substitute**. *DERIVE* will ask us what to substitute for each variable. Press Enter for x and y to indicate that they are to be left unchanged. When asked for a substitute value for z type `-8` Enter. **Simplify** expression 3 in Figure 1.2 to obtain the solution in expression 4. Again, we stress that there are many correct answers to the second part of the problem.

Solved Problem 3: Solve the following system of equations.

$$\begin{aligned} 3x + 4y + z &= 8 \\ 2x - y + 3z &= 2 \\ 3x + 4y + z &= 8 \end{aligned}$$

Solution: Notice that the first two equations are the same as those in Solved Problem 2. Since the third equation is the same as the first, the solution of the system is the same as the one in Solved Problem 2; but as we shall see, *DERIVE* presents it a little differently.

```
5:   [3 x + 4 y + z = 8, 2 x - y + 3 z = 2, 3 x + 4 y + z = 8]
```

$$6: \quad \left[x = @1,\ y = \frac{22 - 7\,@1}{13},\ z = \frac{16 - 11\,@1}{13}\right]$$

$$7: \quad \left[x = x,\ y = \frac{22 - 7\,x}{13},\ z = \frac{16 - 11\,x}{13}\right]$$

$$8: \quad \left[y = \frac{22 - 7\,x}{13},\ z = \frac{16 - 11\,x}{13}\right]$$

Figure 1.3: **Alternate form for the case of many solutions**

Author `[3x+4y+z=8,2x-y+3z=2,3x+4y+z=8]` and ask *DERIVE* to **soLve**. This time *DERIVE* does not ask us for solve variables. It only does that when it cannot determine by itself which variables to solve for. Instead it presents the solution in expression 6 of Figure 1.3 immediately in terms of the funny symbol @1. Don't be confused by its unusual appearance, it's just *DERIVE*'s way of introducing parameters. Think of @1 as "arbitrary 1." We may treat it just as we would any other variable. The reason *DERIVE* doesn't use an ordinary name such as s or t is that it might conflict with an earlier use of it.

We *should* be able to make this solution look like the one for Solved Problem 2. The first step is to replace the variable @1 by x. With expression 6 of Figure 1.3 highlighted, use **Manage Substitute**. Press Enter for x, y, and z to leave them as they are, but replace @1 by x. The solution in expression 2 of Figure 1.2 is presented with solve variables x and y, while expression 7 uses solve variables y and z. Thus, they still look different. To correct this, highlight expression 2 of Figure 1.2 once more and **soLve**. When *DERIVE* asks for solve variables, respond with **y** and **z**. The result is expression 8 of Figure 1.3 and the two solutions finally look the same.

When solving systems of equations we often have choices to make for the solve variables. Different choices may make the answers *look* different, but they are, in fact, equivalent.

Solved Problem 4: ACE milling company has received an order for 1000 pounds of a special feed that contains 4% fat, 15% fiber, and 15% protein. They must mix the required feed using the ingredients they have on hand: wheat mids, [1] rice mill feed, cotton seed meal, soy bean hulls and alfalfa. The following table gives the percentages of fat, fiber, and protein plus the cost per pound for each ingredient.

	% fat	% fiber	% protein	cost per pound ($)
w = wheat mids	3	5	15	0.05
r = rice mill feed	6	30	5	0.02
c = cotton seed meal	3	10	40	0.10
s = soy bean hulls	2	35	10	0.05
a = alfalfa	4	25	20	0.08

How should they mix the feed if the cost is to be a minimum? Give the amounts of each ingredient used and the total cost.

Solution: The first step is to write the constraints on the composition of the feed as a system of equations. Let w, r, c, s, and a denote the total number of pounds of each ingredient (indicated in the table) to be used. Four percent of the 1,000 pounds of feed is to be fat. Thus, from the % fat column in the table we obtain the following equation.

$$3\%w + 6\%r + 3\%c + 2\%s + 4\%a = 40$$

Remark: We may enter a percent into *DERIVE* in this way.

In a similar fashion we obtain the following equations for fiber and protein.

$$5\%w + 30\%r + 10\%c + 35\%s + 25\%a = 150$$

$$15\%w + 5\%r + 40\%c + 10\%s + 20\%a = 150$$

Finally the total amount of feed is 1000 pounds, so that $w + r + c + s + a = 1000$.

Notice that there are four equations in five unknowns, so we do not expect a unique solution. **Author** the system and **soLve**. We can use any four of the variables as solve variables. The solution in expression 2 of Figure 1.4 is obtained using a, c, r, and s as solve variables. *(Note that expressions 1, 4, and 6 are incomplete as they are too long to fit on the screen.)*

[1]When the wheat kernel is removed to produce flour, some parts of the whole wheat including the bran are often pressed into pellets known as "wheat mids" and sold as a feed ingredient.

```
1:    [3% w + 6% r + 3% c + 2% s + 4% a = 40, 5% w + 30% r + 10% c + 35% s + 25%

2:    [a = 4 (3 w - 1475), c = 5 (520 - w), r = 5 (560 - w), s = 1500 - 3 w]

3:    0.05 w + 0.02 r + 0.1 c + 0.05 s + 0.08 a

4:    0.05 w + 0.02 (5 (560 - w)) + 0.1 (5 (520 - w)) + 0.05 (1500 - 3 w) + 0.08

       13 w - 4050
5:    ------------
           50

            ┌   ┌   1475        ┐         ┌        1475 ┐         ┌        1475 ┐
6:    │a = 4 │3 ------ - 1475│, c = 5 │520 - ------│, r = 5 │560 - ------│, s = 1
            └   └     3          ┘         └          3   ┘         └          3   ┘

       ┌                425          1025          ┐
7:    │a = 0, c = ------, r = -------, s = 25│
       └                 3             3           ┘
```

Figure 1.4: **A feed ration of minimal cost**

The next step is to **Author** the cost function as expression 3 in Figure 1.4, then use **Manage Substitute** to replace a by $4(3w - 1475)$, c by $5(520 - w)$, r by $5(560 - w)$ and s by $1500 - 3w$ as seen in expression 4. **Simplify** expression 4 to obtain the cost in expression 5 as a function of w.

The cost, $\dfrac{13w - 4050}{50}$, is clearly minimal if w is chosen as small as possible. But the physical constraints on the problem require that each variable a, c, r and s be non-negative. From expression 2 in Figure 1.4 these constraints give the following inequalities.

$$\begin{aligned} 4(3w - 1475) &\geq 0 \\ 5(520 - w) &\geq 0 \\ 5(560 - w) &\geq 0 \\ 1500 - 3w &\geq 0 \end{aligned}$$

We want to find the *smallest* value of w that makes all four of these inequalities true. If we solve each inequality and compare the solutions, we find that $w = \dfrac{1475}{3}$. It is probably faster to do this by hand than it is to enter all the data into *DERIVE*.

Use **Manage Substitute** to replace w in expression 2 by $\frac{1475}{3}$ and **Simplify** to obtain the amounts of the other ingredients in expression 7 of Figure 1.4. If we put $w = \dfrac{1475}{3}$ into the cost function in expression 5, we obtain the minimum cost of \$46.83 for the 1000 pounds of feed.

Exercises

1. Solve the following systems of equations. If the solution is not unique, find a specific solution with x between 20 and 30.

   (a)

$$\begin{aligned} 4x + 6y - z + 8w &= 5 \\ 3x + 6y - 4z + w &= 8 \\ 5x - y + 7z - 8w &= 3 \end{aligned}$$

   (b)

$$\begin{aligned} 6x - 2y + 8z + 4w &= 5 \\ 5x - 4y - 3z + 2w &= 10 \\ 2x - 7y - z - w &= -3 \\ x + 2y + 11z + 2w &= -5 \end{aligned}$$

   (c)

$$\begin{aligned} 2x + y + 3z - w &= 8 \\ x - y + 2z + 2w &= 4 \\ 3x + 5z + w &= 12 \end{aligned}$$

   (d)

$$\begin{aligned} x + y + z + w &= 8 \\ 2x + 3y - z + w &= 4 \\ 3x + 4y + 2w &= 5 \end{aligned}$$

2. Solve the following system of equations for x, y, and z.

$$\begin{aligned} ax + 3y + 7z &= 9 \\ 2x + 2y - z &= 4 \\ cx - y + 5z &= 2 \end{aligned}$$

   What is the solution if $a = 3$ and $c = 7$?

   What is the solution if $a = \frac{61}{9}$ and $c = 1$?

3. A bag contains 100 coins consisting of nickels, dimes and quarters. There is a total of \$22.00 in the bag.

   (a) What is the maximum possible number of quarters?

   (b) What is the minimum possible number of quarters?

   (c) List all possibilities for the distribution of coins in the bag.
   (Hint: There are six possibilities.)

4. Suppose that *neither* of the following systems of equations has a solution.

$$\begin{aligned} ax + 3y - z &= 2 \\ bx + 2y - 3z &= 5 \\ 2x - y + 3z &= 4 \end{aligned} \qquad \begin{aligned} ax - 2y + 7z &= 9 \\ bx + 5y + 2z &= 6 \\ 4x + 7y - 8z &= 1 \end{aligned}$$

   Find a and b.

5. The following problems are a continuation of Solved Problem 4.

   (a) Find the maximum possible percentage of fat that the feed mixture can have if the other restrictions on the feed remain the same.

   (b) Suppose that the required percentage of fat is 5%. Give the amounts of each ingredient so that the feed is of minimum cost and find this cost.

   (c) Suppose that you find yourself overstocked with alfalfa. Rather than produce a mixture of minimal cost, you wish to fill the order using a maximum amount of alfalfa. Find the amounts of each ingredient and the total cost of this alfalfa-rich feed.

6. A manufacturer wishes to melt and combine some or all of six alloys that are available from a distributor to produce 100 pounds of a new alloy that is (by weight) 6% gold, 15% silver, and 30% iron. The five available alloys and their percentage content of gold, silver, and iron are given in the table below.

Alloy	% Gold	% Silver	% Iron
A	12	10	35
B	6	23	15
C	16	5	15
D	5	7	42
E	2	28	17
F	4	8	55

Prepare an order for the distributor that will allow you to produce the required alloy.

7. A grain silo typically has an irregularly shaped hopper at the base. At inventory time, someone stands at the top of the silo and drops a tape down the interior to the surface of the grain. The reading (in feet) is reported to the manager who consults a table (provided by the builder) that gives the number of cubic feet of grain remaining in the silo based on the distance from the surface of the grain to the top of the silo. An actual table for a silo owned by Stillwater Milling Company is given below.

depth	cubic feet	depth	cubic feet	depth	cubic feet	depth	cubic feet
0	2815	13	1983	26	1151	39	319
1	2751	14	1919	27	1087	40	255
2	2687	15	1855	28	1023	41	196.5
3	2623	16	1791	29	959	42	147.5
4	2559	17	1727	30	895	43	106.9
5	2495	18	1663	31	831	44	74.9
6	2431	19	1599	32	767	45	49.5
7	2367	20	1535	33	703	46	30.5
8	2303	21	1471	34	639	47	16.5
9	2339	22	1407	35	575	48	7.2
10	2715	23	1343	36	511	49	1.8
11	2111	24	1279	37	447	50	0
12	2047	25	1215	38	383		

We want to find a function $C(x)$ that gives the cubic feet of grain in terms of the distance x from the surface of the grain to the top of the silo. The data points 0 through 40 correspond to the regular cylindrical portion of the silo, and thus are described by a linear function $f(x) = ax + b$ for a suitable choice of a and b. The last 10 data points for depths 41 through 50 cannot be accurately approximated by a linear function, so the idea is to find a polynomial function $g(x)$ that does. When we do, we get

$$C(x) = \begin{cases} f(x) & \text{if } x \leq 40 \\ g(x) & \text{if } 40 < x \leq 50 \end{cases}$$

(a) Find $f(x)$.

(b) There are a number of ways to find a function $g(x)$ that approximates the last 10 data points. (We shall return to this problem once more when we look at least squares approximations.) We will use a fourth-degree polynomial that passes through the data points 41, 43, 45, 47, and 50, and check that the function doesn't miss the remaining points at depths 42, 44, 46, 48, and 49 feet by very much.

If $g(x) = ax^4 + bx^3 + cx^2 + dx + e$ fits the data points for depths 41, 43, 45, 47, and 50 feet, write a system of five equations in five unknowns that must be solved to determine the coefficients a, b, c, d, and e.

(c) Find $g(x)$.

(d) For depths 42, 44, 46, and 48 feet, calculate $g(x)$ and the error.

(e) Find the approximate amount of grain in the silo when the depth is 44.5 feet.

Exploration and Discovery

1. Ask *DERIVE* to solve the following system of equations for x and y.

$$\begin{aligned} ax + by &= c \\ dx + ey &= f \end{aligned}$$

   Is this solution correct for all values of a, b, d and e? If not, analyze the solution in cases where *DERIVE*'s solution is not correct and discuss your findings.

2. Solve the following system of equations using x and y as solve variables.

$$\begin{aligned} x + y + 2z + 2w &= 3 \\ 2x + y + 3z + 3w &= 4 \end{aligned}$$

   Ask *DERIVE* to solve the system once more using z and w as solve variables. Explain what happens. In general, which variables in a system of linear equations can be chosen to be solve variables?

3. An equation of the form $ax + by + cz = d$ is the equation of a plane, and the solution of a system of 3 equations in three unknowns is the common intersection of three planes. *DERIVE* can be made to produce a nice picture of this intersection using the `MAX` function. (For example, if you **Author** and **Simplify** the expression `max(3,7,2,4)`, *DERIVE* will return the largest of the numbers in the list; 7 in this case.)

   Solve the following system of equations.

$$\begin{aligned} x + y + z &= 3 \\ x + y - z &= 1 \\ x - y + z &= 2 \end{aligned}$$

   To get *DERIVE* to plot the three planes we need to solve each equation for z.

$$\begin{aligned} z &= 3 - x - y \\ z &= x + y - 1 \\ z &= 2 - x + y \end{aligned}$$

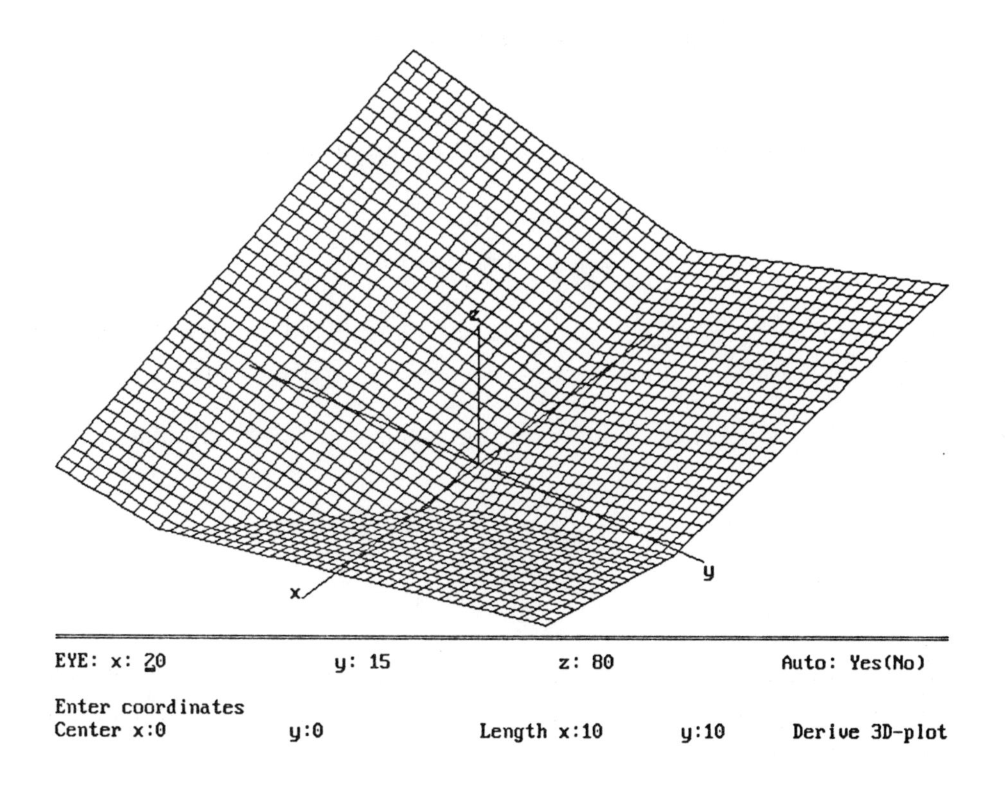

Figure 1.5: **The intersection of three planes**

Author the expression `max(3-x-y,x+y-1,2-x+y)` and then execute the **Plot** command twice. *DERIVE* will produce the picture in Figure 1.5.

You may wish to increase the number of **Grid** lines to 40 to produce a nice picture. (For further instructions on plotting, consult Section 5.8 of your *DERIVE* manual.)

From a geometric point of view, what can you say about the solutions of systems of three equations in three unknowns in case the solution is not unique?

LABORATORY EXERCISE 1.1

Producing a Dietary Supplement

Name ______________________ **Due Date** ______________

The problem: You wish to produce 1000 kilograms of a product containing 0.25% folic acid, 0.25% vitamin B-6, 12% calcium, and 2% iron. This product is to be pressed into tablets and sold as a dietary supplement. The tablets are to be made from some (or all) of six preparations that can be obtained from your distributor. The pertinent information is displayed in the following table.

	%folic acid	%vitamin B-6	%calcium	%iron	cost($ per kilogram)
A	0.1	0.1	18	3	4.20
B	0.3	0	9	3	3.70
C	0.2	0.4	14	1	2.50
D	0.1	0.5	19	1.2	5.60
E	0.4	0.2	10	2	7.20
F	0.2	0.2	11	3	4.50

1. Write a system of equations whose solution provides the appropriate amount of each ingredient.

continued on next page

2. Solve the system of equations.

3. Express the cost in terms of a single variable.

4. What value of the variable used in Part 3 gives the minimum cost? Explain carefully how you arrived at this value.

5. How much of each ingredient should you use to obtain a product of minimum cost?

6. If you press the product into 4-gram capsules and sell it in bottles of 250 for $6.00 per bottle, what is your profit?

CHAPTER 2

AUGMENTED MATRICES AND ELEMENTARY ROW OPERATIONS

LINEAR ALGEBRA CONCEPTS

- **Augmented matrix**
- **Elementary row operation**
- **Echelon form**
- **Reduced echelon form**
- **Gauss-Jordan Elimination**
- **Transpose**
- **Rank (optional)**

Introduction

A matrix is said to be in *echelon form* if (1) the first nonzero entry in a nonzero row is "1," (2) all zero rows are below all nonzero rows, and (3) the leading "1" in a nonzero row occurs to the right of the leading "1" in any previous row. If, in addition, each column with a leading "1" has zeros everywhere else, then the matrix is said to be in *reduced echelon form* (or simply *row reduced*). *DERIVE* 's internal function ROW_REDUCE returns the reduced echelon form of a matrix.

In several problems we will find it useful to load auxiliary files. VECTOR.MTH comes with the *DERIVE* program, but REDUCE.MTH must be entered and saved by the user. It is found in Appendix II.

DERIVE 's syntax for the transpose of A is A‘. That's *not* an apostrophe (') – it's located on the same key as ˜ on a standard keyboard.

Solved Problems

Solved Problem 1: **This exercise uses the file VECTOR.MTH** Display the row operations required to put $\begin{bmatrix} 0 & 3 & 5 \\ 2 & 4 & 0 \\ 4 & 8 & 1 \end{bmatrix}$ into echelon form.

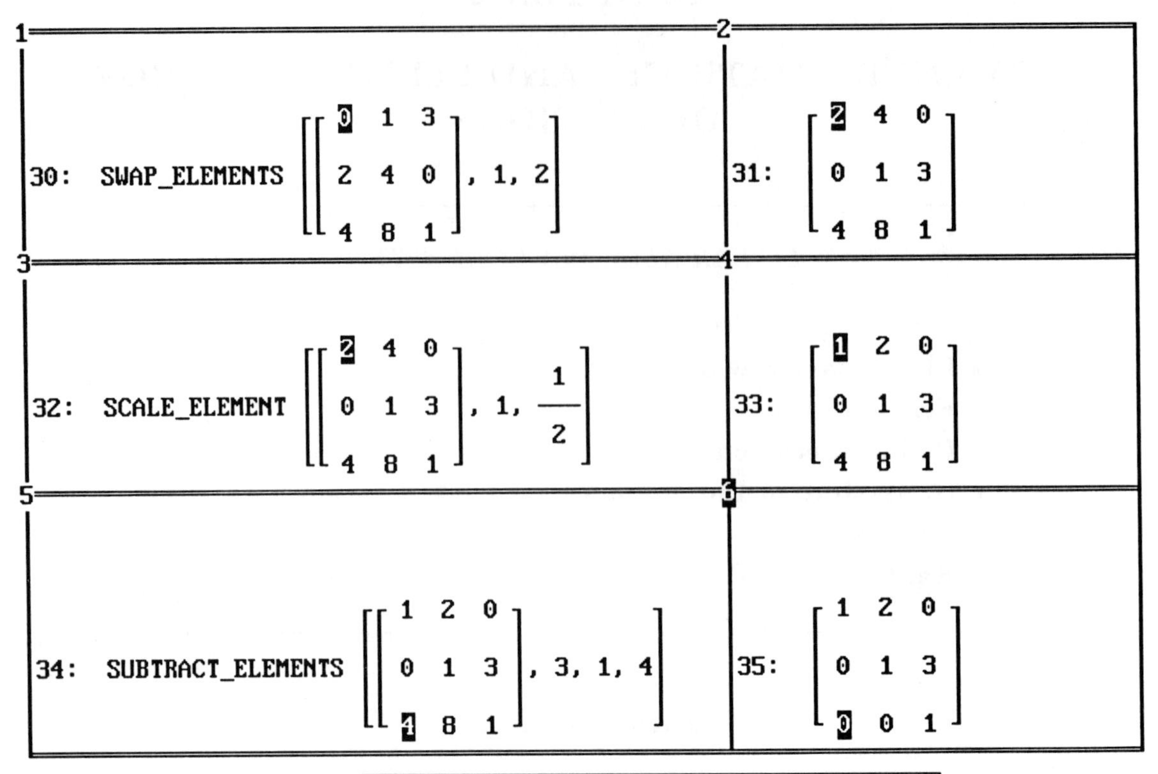

Figure 2.1: **Step by step row reduction of a matrix**

Solution: Each of the three elementary row operations is included in the VECTOR.MTH file that comes with the *DERIVE* program.

First, use **Transfer Merge** and type the file name VECTOR.

Row Operations from the VECTOR.MTH File

SCALE_ELEMENT(A,r,s) multiplies row r of matrix A by the number s.

SUBTRACT_ELEMENTS(A,i,j,s) subtracts s times row j of matrix A from row i.

SWAP_ELEMENTS(A,i,j) swaps rows i and j of matrix A.

To enter a matrix into *DERIVE*, issue the **Declare Matrix** commands. *DERIVE* will ask you for the size of the matrix you wish to enter. The defaults of 3 rows and columns are correct, so press [Enter]. *DERIVE* will now prompt you to enter the elements of the matrix. For details see Section 1 in Appendix I.

To swap rows 1 and 2, highlight the matrix A and **Author** `swap_elements(`[F3]`,1,2)`. (Alternatively, you may refer to the matrix by its expression number: Assuming it is expression 29, **Author** `swap_elements(#29,1,2)`.) **Simplify** to obtain expression 31 of Figure 2.1. (For economy of presentation we have split the window in Figure 2.1 several times. It is not necessary for you to do this, but if you wish to learn more about window splitting, consult Section 5.7 of your *DERIVE* manual.)

The next step is to multiply row 1 by $\frac{1}{2}$ using `scale_element(`[F3]`,1,1/2)`. To subtract 4 times row 1 from row 3, **Author** `subtract_elements(`[F3]`,3,1,4)`. An echelon form appears in expression 35 of Figure 2.1. (You should be aware that the echelon form of a matrix is not unique. Thus if you perform a different sequence of row operations, you may arrive at a correct answer that is different from that in Figure 2.1.)

A *DERIVE* Suggestion: Some users, like us, may find it convenient to rename some of these functions to shorten their names, and even change their arguments to make them more consistent with the way elementary row operations are presented in most texts. For example, if you **Author** `add(s,j,i,a):=subtract_elements(a,i,j,-s)` then ADD(s,j,i,a) will add s times row j of the matrix A to row i. (You may save this in a file using **Transfer Save** for later use if you wish.) *To keep this book consistent with the DERIVE manual, we will use the DERIVE Manual's definitions.*

Solved Problem 2: Determine if it is possible by elementary row operations to transform the matrix $\begin{bmatrix} 1 & 2 & 3 \\ 4 & 5 & 6 \\ 7 & 8 & 9 \end{bmatrix}$ into the matrix $\begin{bmatrix} 2 & 5 & 2 \\ 1 & 4 & 3 \\ 3 & 2 & 1 \end{bmatrix}$.

Solution: Enter the first matrix as described in Solved Problem 1 above. Next **Author** and **Simplify** `row_reduce`[F3]. The reduced echelon form is expression 3 of Figure 2.2. Repeat this procedure to get the reduced echelon form of the second matrix (expression 6). The two are obviously different. Since the reduced echelon form is unique, we conclude that it is not possible to obtain the second matrix from the first using elementary row operations.

DERIVE hint: If you intend to do more than one operation on a given matrix, it is helpful to give it a name. For example, if you highlight expression 1 of Figure 2.2 and **Author** the expression `A:=`[F3], *DERIVE* will know that the name A refers to the matrix in expression 1. You can now get the row-reduced form of A using `row_reduce(A)`.

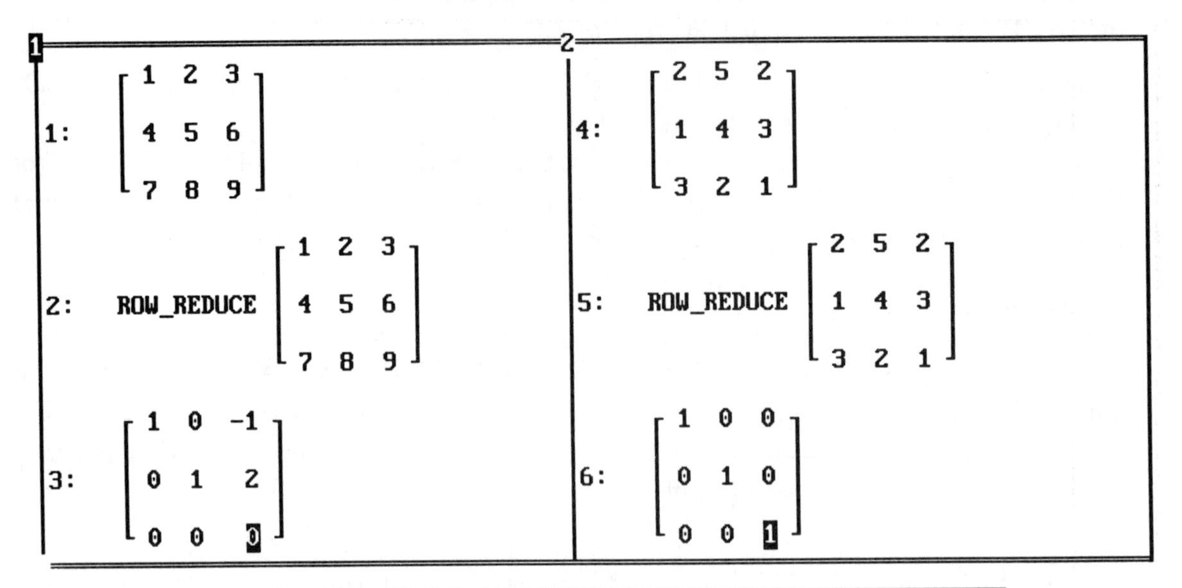

Figure 2.2: Matrices with different reduced echelon forms.

Solved Problem 3: Use Gauss-Jordan elimination to solve the following two systems.

$$\begin{aligned} 2x + 3y - 7z + w &= 8 \\ 4x - 2y + z + 2w &= 4 \\ 5x + y - 4z + 3w &= 6 \end{aligned} \qquad \begin{aligned} 2x + 3y - 7z + w &= 8 \\ 4x - 2y + z + 2w &= 4 \\ 6x + y - 6z + 3w &= 5 \end{aligned}$$

Solution: The augmented matrix of the first system is

$$\left[\begin{array}{rrrr|r} 2 & 3 & -7 & 1 & 8 \\ 4 & -2 & 1 & 2 & 4 \\ 5 & 1 & -4 & 3 & 6 \end{array}\right]$$

To enter this matrix, issue the **Declare Matrix** commands just as in Solved Problem 1. *DERIVE* will ask you for the size of the matrix you wish to enter. The default of 3 rows is correct, but the number of columns must be changed to 5. Use the Tab key to move the cursor to **Columns:**, change the 3 to 5, and press Enter. *DERIVE* will now prompt you to enter the elements of the matrix. (For more details, see Section 1 in Appendix I.) When you have finished, the matrix will appear as seen in expression 1 of Figure 2.3.

Author and **Simplify** `row_reduce(#1)`. The solution of the system of equations can now be read from expression 3 of Figure 2.3 by solving for the leading variables.

1:
$$\begin{bmatrix} 2 & 3 & -7 & 1 & 8 \\ 4 & -2 & 1 & 2 & 4 \\ 5 & 1 & -4 & 3 & 6 \end{bmatrix}$$

2: ROW_REDUCE
$$\begin{bmatrix} 2 & 3 & -7 & 1 & 8 \\ 4 & -2 & 1 & 2 & 4 \\ 5 & 1 & -4 & 3 & 6 \end{bmatrix}$$

3:
$$\begin{bmatrix} 1 & 0 & 0 & \frac{16}{21} & -\frac{10}{21} \\ 0 & 1 & 0 & \frac{5}{7} & -\frac{32}{7} \\ 0 & 0 & 1 & \frac{8}{21} & -\frac{68}{21} \end{bmatrix}$$

4:
$$\begin{bmatrix} 2 & 3 & -7 & 1 & 8 \\ 4 & -2 & 1 & 2 & 4 \\ 6 & 1 & -6 & 3 & 5 \end{bmatrix}$$

5: ROW_REDUCE
$$\begin{bmatrix} 2 & 3 & -7 & 1 & 8 \\ 4 & -2 & 1 & 2 & 4 \\ 6 & 1 & -6 & 3 & 5 \end{bmatrix}$$

6:
$$\begin{bmatrix} 1 & 0 & -\frac{11}{16} & \frac{1}{2} & 0 \\ 0 & 1 & -\frac{15}{8} & 0 & 0 \\ 0 & 0 & 0 & 0 & 1 \end{bmatrix}$$

Figure 2.3: **Solving by Gauss-Jordan elimination**

$$\begin{aligned} x &= -\frac{10}{21} - \frac{16}{21}w \\ y &= -\frac{32}{7} - \frac{5}{7}w \\ z &= -\frac{68}{21} - \frac{8}{21}w \\ w & \quad \text{is arbitrary} \end{aligned}$$

To solve the second system of equations, follow the instructions above to enter the augmented matrix $\begin{bmatrix} 2 & 3 & -7 & 1 & 8 \\ 4 & -2 & 1 & 2 & 4 \\ 6 & 1 & -6 & 3 & 5 \end{bmatrix}$. Now **Author** and **Simplify** `row_reduce(#4)`. From the last row of the matrix in expression 6 of Figure 2.3, we conclude that this system of equations has no solution.

You may wish to compare your work here with what happens if you solve these systems using the methods of Chapter 1.

Solved Problem 4: **This exercise uses the file REDUCE.MTH. See Appendix II.** Find values of p, q, and r so that the following system of equations has no solution.

$$\begin{aligned} 2x + y + 3z + 5w &= p \\ 3x + 2y - z + 4w &= q \\ 3x + y + 10z + 11w &= r \end{aligned}$$

$$10: \quad \begin{bmatrix} 2 & 1 & 3 & 5 & p \\ 3 & 2 & -1 & 4 & q \\ 3 & 1 & 10 & 11 & r \end{bmatrix}$$

$$11: \quad \text{REDUCE}\left[\begin{bmatrix} 2 & 1 & 3 & 5 & p \\ 3 & 2 & -1 & 4 & q \\ 3 & 1 & 10 & 11 & r \end{bmatrix}, 4\right]$$

$$12: \quad \begin{bmatrix} 1 & 0 & 7 & 6 & \frac{2r}{3} - \frac{q}{3} \\ 0 & 1 & -11 & -7 & q - r \\ 0 & 0 & 0 & 0 & p - \frac{q}{3} - \frac{r}{3} \end{bmatrix}$$

Figure 2.4: **Making a system of equations inconsistent**

Solution: If you have not already loaded the REDUCE.MTH file, do so now using **Transfer Load Utility** REDUCE. Enter the 3 by 5 augmented matrix of this system using **Declare Matrix** as described in the solved problems above. Next, assuming the matrix is expression 10, **Author** and **Simplify** `reduce(#10,4)`. From the last row of the matrix in expression 12 of Figure 2.4, we see that the system of equations has a solution if and only if $p - \frac{q}{3} - \frac{r}{3} = 0$. Thus, any choice of p, q, and r that does not satisfy this equation will work. One correct solution is $p = q = r = 1$. (See what happens if you use ROW_REDUCE rather than the REDUCE function.)

Exercises

1. Use Gauss-Jordan Elimination to solve the following systems.

(a)
$$\begin{aligned} 2x - 3y + 4z - 5w &= 6 \\ 3x + 4y - 2z + w &= 4 \\ 5x - 2y + z + 3w &= 8 \end{aligned}$$

(b)
$$\begin{aligned} 3a + 4b - 7c + 6d - e &= 1 \\ 4a + b - 2c + 4d - 3e &= 4 \\ 2a + 3b + c + 4d + 7e &= 5 \end{aligned}$$

(c)
$$\begin{aligned} 5x + 4y - z + w &= 8 \\ 2x - y + 2z - w &= 3 \\ 7x + 3y + z &= 2 \end{aligned}$$

2. **This exercise uses the file** REDUCE.MTH. See Appendix II. Find values of p, q, r, and s so that the following system of equations has no solution.

$$\begin{aligned} 3a + 4b + 5c - d + e &= p \\ 4a + 2b - c + 7d - e &= q \\ 2a - 3b + 4c + d + 3e &= r \\ 5a + 9b + 5d - 3e &= s \end{aligned}$$

3. **This exercise uses the file REDUCE.MTH. See Appendix II.** Each of the following three systems of equations has a solution. Find a, b, and c.

$$\begin{aligned} 2x+3y+4z &= a \\ 4x+2y-z &= b \\ 6x+5y+3z &= c \end{aligned} \qquad \begin{aligned} 3x-2y+7z &= a \\ 2x-4y+6z &= b \\ 4x+8z &= c \end{aligned} \qquad \begin{aligned} 5x+4y-9z &= a \\ 3x+6y+z &= b \\ 2x-2y-10z &= c \end{aligned}$$

4. Solve the following systems of equations.

(a)
$$\begin{aligned} 2x^3+5y^3-7z^3+w^3 &= 5 \\ 4x^3-3y^3+z^3-2w^3 &= 2 \\ 3x^3+3y^3+5z^3-w^3 &= 1 \\ 6x^3-5y^3+4z^3+9w^3 &= 7 \end{aligned}$$

(b)
$$\begin{aligned} (a+b+c)^2+5(a+b+c)-2 &= 0 \\ 3a-2b+7c &= 0 \\ 5a+4b+2c &= 9 \end{aligned}$$

(Hint: *DERIVE* will not solve this directly. Consider solving $x^2+5x-2=0$ first.)

5. Determine if $\begin{bmatrix} 2 & 3 & 4 & 5 \\ 8 & 2 & 5 & 1 \\ 7 & 7 & 5 & 3 \end{bmatrix}$ can be transformed into $\begin{bmatrix} 3 & 6 & 9 & 2 \\ 4 & 5 & 0 & 1 \\ 3 & 1 & 5 & 8 \end{bmatrix}$ by elementary row operations.

6. Find a solution of the following system of equations such that x, y, z and w are integers.

$$\begin{aligned} 2x+y-z+w &= 5 \\ 3x-2y+z-w &= 0 \\ 2x+2y-2z+w &= 3 \end{aligned}$$

7. Let $A = \begin{bmatrix} 1 & 2 & 3 & 2 \\ 4 & 5 & 6 & 11 \\ 7 & 8 & 9 & 10 \end{bmatrix}$ and $B = \begin{bmatrix} 1 & 2 & 3 & 1 \\ 4 & 5 & 6 & 11 \\ 7 & 8 & 9 & 10 \end{bmatrix}$.

(a) Find the transposes of the row-reduced forms of A and B. (*DERIVE* 's syntax for the transpose of A is A‘.)

(b) Find the row-reduced forms of the transposes of A and B.

(c) What does this say about the proposition that the reduced form of the transpose is the transpose of the reduced form?

Exploration and Discovery

1. For this exercise we will define the *rank* of a matrix M to be the number of non-zero rows in the reduced echelon form of M. It is easy to calculate using *DERIVE* – just ROW_REDUCE and count the number of non-zero rows. For example, if $M = \begin{bmatrix} 1 & 2 & 3 \\ 4 & 5 & 6 \\ 7 & 8 & 9 \end{bmatrix}$ then the reduced echelon form of M is $\begin{bmatrix} 1 & 0 & -1 \\ 0 & 1 & 2 \\ 0 & 0 & 0 \end{bmatrix}$, so the rank of M is 2. (Other equivalent definitions of "rank" are discussed further in Chapter 8.)

   (a) Do Exercise 7.

   (b) Let A and B be as in Exercise 7. Calculate the ranks of A and B.

   (c) Calculate the ranks of the transposes of A, B, and M above.

   (d) Experiment with other matrices of your choice, including a 3 by 5 and a 4 by 2. Discuss your observations and conclusions. (See **The Random Matrix File** in Appendix II for instructions on how to generate random examples.)

   (e) Suppose that A is a matrix whose rank is equal to the number of rows of A. What can you say about the consistency of a system of equations whose augmented matrix is $\left[\begin{array}{c|c} A & B \end{array}\right]$?

2. **Maximizing a linear function over a convex polygon.** In many applications it is important to maximize a linear function subject to constraints introduced by a system of linear inequalities. Such problems are referred to as "linear programming" problems. For example, let's maximize the function $C(x, y) = 2x + 3y - 1$ subject to the following system of inequalities:

$$\begin{aligned} x &\geq 0 \\ y &\geq 0 \\ x + y &\leq 1 \end{aligned}$$

Each inequality describes a half-plane (see Figure 2.5). For example, the inequality $x+y \leq 1$ describes the half-plane below the line $x + y = 1$. Linear inequalities taken together describe an intersection of half-planes that is often a convex polygon. The inequalities above, for example, describe the triangle in the plane with vertices $(0,0)$, $(0,1)$, and $(1,0)$. **It is known that the maximum of the linear function $C(x,y)$ must occur at a vertex of the polygon, perhaps including an edge as well.** Since $C(0,0) = -1$, $C(1,0)$=1 and $C(0,1) = 2$, we conclude that, subject to the constraints above, C achieves its maximum, 2, at the point $(0,1)$.

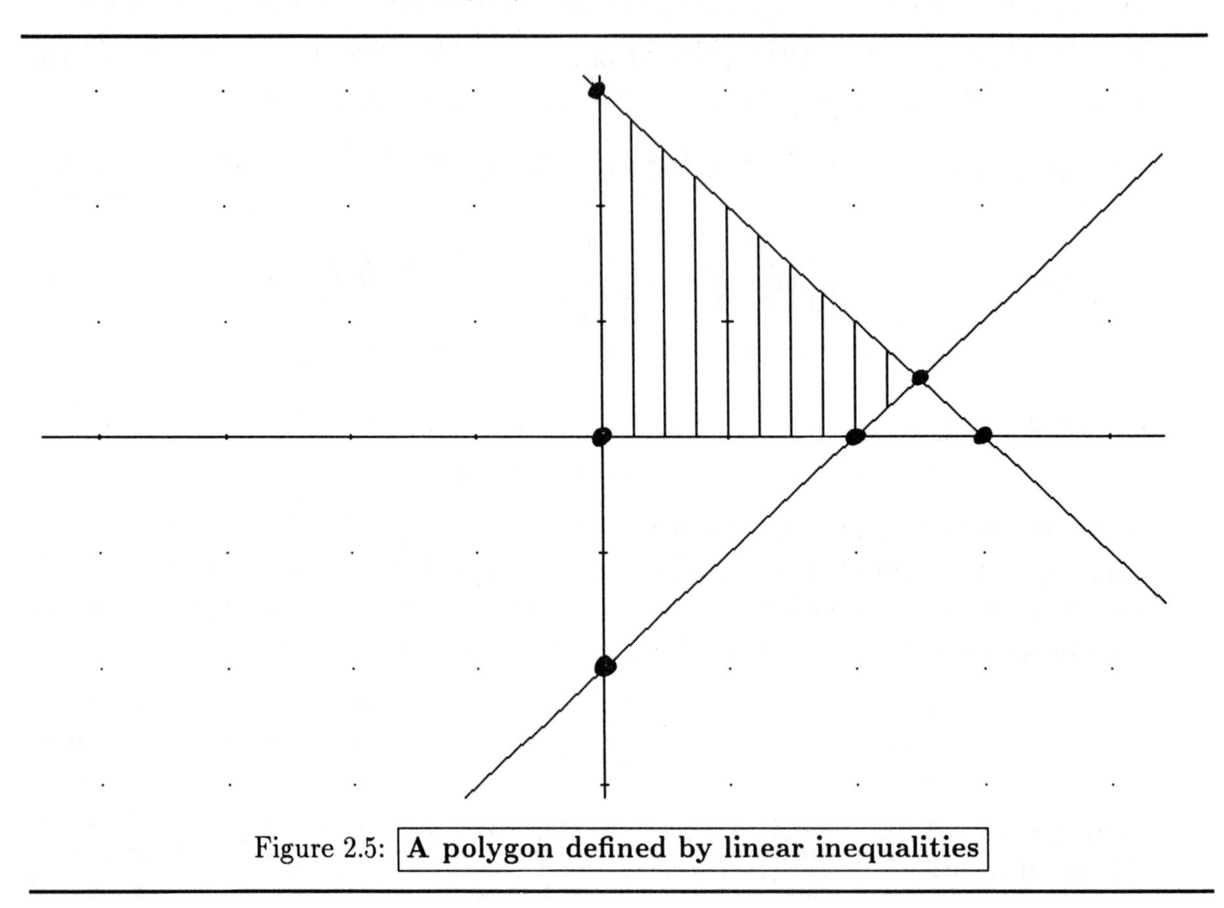

Figure 2.5: **A polygon defined by linear inequalities**

In general, the difficulty in solving a problem like the one above is in finding the vertices of the polygon. *DERIVE* can help. Let's maximize $C(x,y) = 2x + 3y - 1$ above subject to the following constraints.

$$\begin{aligned} x &\geq 0 \\ y &\geq 0 \\ x+y &\leq 3 \\ x-y &\leq 2 \end{aligned}$$

The vertices of the convex region are among the intersections of the lines $x = 0$, $y = 0$, $x + y = 3$, and $x - y = 2$. Thus we need to solve the following six systems of equations.

$$\begin{aligned} x &= 0 \\ y &= 0 \end{aligned} \qquad \begin{aligned} x &= 0 \\ x+y &= 3 \end{aligned} \qquad \begin{aligned} x &= 0 \\ x-y &= 2 \end{aligned}$$

$$\begin{aligned} y &= 0 \\ x+y &= 3 \end{aligned} \qquad \begin{aligned} y &= 0 \\ x-y &= 2 \end{aligned} \qquad \begin{aligned} x+y &= 3 \\ x-y &= 2 \end{aligned}$$

Enter the 4 by 3 augmented matrix of the system of inequalities $\begin{bmatrix} 1 & 0 & 0 \\ 0 & 1 & 0 \\ 1 & 1 & 3 \\ 1 & -1 & 2 \end{bmatrix}$. It will be convenient to give this matrix a name. **Author** `A:=` [F3]. *DERIVE* now knows that A refers to the augmented matrix above. Notice that the augmented matrix of any of the six systems of two equations in two unknowns above can be obtained by selecting two appropriate rows from A. For example the augmented matrix of

$$\begin{aligned} x &= 0 \\ x+y &= 3 \end{aligned}$$

consists of rows 1 and 3 of A. We need an efficient way to get the solutions of the six systems of equations above. The following function will help. **Author**

```
f(i,j):=row_reduce([element(a,i),element(a,j)])
```

The function $f(i,j)$ returns the reduced form of the matrix consisting of rows i and j of A. To obtain the solutions of the six systems of equations above, **Author** and **Simplify** the six expressions: `f(1,2)`, `f(1,3)`, `f(1,4)`, `f(2,3)`, `f(2,4)`, and `f(3,4)`. *DERIVE* will return reduced augmented matrices from which the following solutions can be read:

$$(0,0),(0,3),(0,-2),(3,0),(2,0),(\tfrac{5}{2},\tfrac{1}{2})$$

Not all of these points are vertices of the polygon. (See Figure 2.5.) Only the points $(0,0)$, $(0,3)$, $(2,0)$, and $(\frac{5}{2},\frac{1}{2})$ satisfy *all four* of the inequalities above. (We find this out by simply checking each of the points in the four inequalities to see if they work.) Finally, $C(0,0)=-1$, $C(0,3)=8$, $C(2,0)=3$, and $C(\frac{5}{2},\frac{1}{2})=\frac{11}{2}$. We conclude that C reaches its maximum value, 8, at the point $(0,3)$.

Find the maximum value of the function $C(x,y)$ above subject to the following inequalities.

$$\begin{aligned} x &\geq 0 \\ y &\geq 0 \\ -x+y &\leq 3 \\ x+y &\leq 5 \\ -2x+y &\geq -4 \end{aligned}$$

LABORATORY EXERCISE 2.1

Row Operations

Name ______________________ **Due Date** ______________________

1. Display the row operations required to put $\begin{bmatrix} 0 & -1 & 0 & 4 & 3 \\ 1 & 5 & -2 & 6 & 5 \\ 8 & 1 & 4 & 7 & 7 \end{bmatrix}$ into echelon form, as we did in Solved Problem 1.

2. Determine if the matrix from part 1 can be transformed into $\begin{bmatrix} 32 & 3 & 16 & 32 & 31 \\ 97 & 16 & 46 & 94 & 92 \\ 72 & 7 & 36 & 71 & 69 \end{bmatrix}$ by elementary row operations. Explain.

3. Determine if the matrix from part 1 can be transformed into $\begin{bmatrix} 97 & 16 & 46 & 94 & 92 \\ 32 & 3 & 16 & 32 & 31 \\ 72 & 17 & 36 & 71 & 69 \end{bmatrix}$ by elementary row operations. Explain.

LABORATORY EXERCISE 2.[illegible]

Row Operations

Name [illegible] Date [illegible]

[illegible]

1. Indicate the row operations required to put [illegible] into echelon form, as we [illegible]

[illegible]

[illegible]

[illegible]

LABORATORY EXERCISE 2.2

Maximizing a Linear Function over a Polygon

Name ______________________________ **Due Date** ____________________

The problem: Find the maximum of the function $C(x,y,z) = 3x + y + 3z$ over the polygon determined by the following inequalities.

$$\begin{aligned} x &\geq 0 \\ y &\geq 0 \\ z &\geq 0 \\ 2x + y + z &\leq 2 \\ x + 2y + 3z &\leq 5 \\ 2x + 2y + z &\leq 6 \end{aligned}$$

Discussion: This problem is similar to Exploration and Discovery Problem 2 except that it is set in one higher dimension. Here the convex polygon is the intersection of half-spaces defined by the inequalities, and the vertices are intersections of triples of planes. In Exploration and Discovery Problem 2 we defined a function `f(i,j)` to select pairs of rows of the augmented matrix and row-reduce them. Here we need to solve systems of 3 equations in 3 unknowns.

Thus we need to select triples of rows from $A = \begin{bmatrix} 1 & 0 & 0 & 0 \\ 0 & 1 & 0 & 0 \\ 0 & 0 & 1 & 0 \\ 2 & 1 & 1 & 2 \\ 1 & 2 & 3 & 5 \\ 2 & 2 & 1 & 6 \end{bmatrix}$. For that, another variable must be added. Define the new function as follows:

```
f(i,j,k):=row_reduce([element(a,i),element(a,j),element(a,k)])
```

continued on next page

1. Find the intersection of each triple of planes defined by the following equations.

$$
\begin{aligned}
x &= 0 \\
y &= 0 \\
z &= 0 \\
2x + y + z &= 2 \\
x + 2y + 3z &= 5 \\
2x + 2y + z &= 6
\end{aligned}
$$

continued on next page

2. Which of the points found in part 1 above are vertices of the polygon defined by the given inequalities?

3. Find the maximum of $C(x, y, z)$ subject to the constraints above.

4. You are to maximize a linear function of 100 variables subject to 500 inequalities each involving 100 variables. Thus, in order to find the vertices of the polygon, you must solve many systems of 100 equations in 100 unknowns. The total number of such systems is given by $\dfrac{500!}{100!400!}$, which in *DERIVE* syntax is COMB(500,100).
Suppose that your computer can solve a system of 100 equations in 100 unknowns in 5 seconds. If you begin your calculations today, when will this project be completed? (Give the month, day, and year.)

CHAPTER 3

THE ALGEBRA OF MATRICES

LINEAR ALGEBRA CONCEPTS

- **Matrix addition**
- **Scalar multiplication**
- **Matrix multiplication**

Introduction

Matrices are far more important in mathematics than simply serving as an aid in solving systems of linear equations. They are studied as objects in their own right with a natural algebraic structure similar in many ways to that of real numbers, but there are also striking differences that we will explore.

Important *DERIVE* Note: We usually don't distinguish between the 1 by 3 matrix [1 2 3] and the vector (1, 2, 3), but *DERIVE* does. You should consult Section 4 in Appendix I for details before proceeding.

Solved Problems

Solved Problem 1: If $A = \begin{bmatrix} 1 & 2 & 3 \\ 2 & 3 & 0 \\ 4 & 8 & 5 \end{bmatrix}$ and $B = \begin{bmatrix} 3 & 8 & -2 \\ 4 & 7 & -1 \\ 0 & 3 & 5 \end{bmatrix}$, calculate $3BA - 4A^3$.

Solution: The first step is to enter the matrix A using **Declare Matrix**. Once the matrix is entered, it will be convenient to give it a name. **Author** `A:=` F3. The ":=" that occurs in expression 2 of Figure 3.1 is *DERIVE*'s syntax for a *definition* rather than an equation. *DERIVE* now knows that the name A refers to the matrix you have just entered, and this definition will remain in effect until you change it. Use the same technique to enter the second matrix and name it B. (See Section 2 in Appendix I concerning uppercase and lowercase letters.)

DERIVE uses the period to denote matrix multiplication. Thus, we **Author** and **Simplify** `3B.A-4A^3`. The result is expression 6 of Figure 3.1.

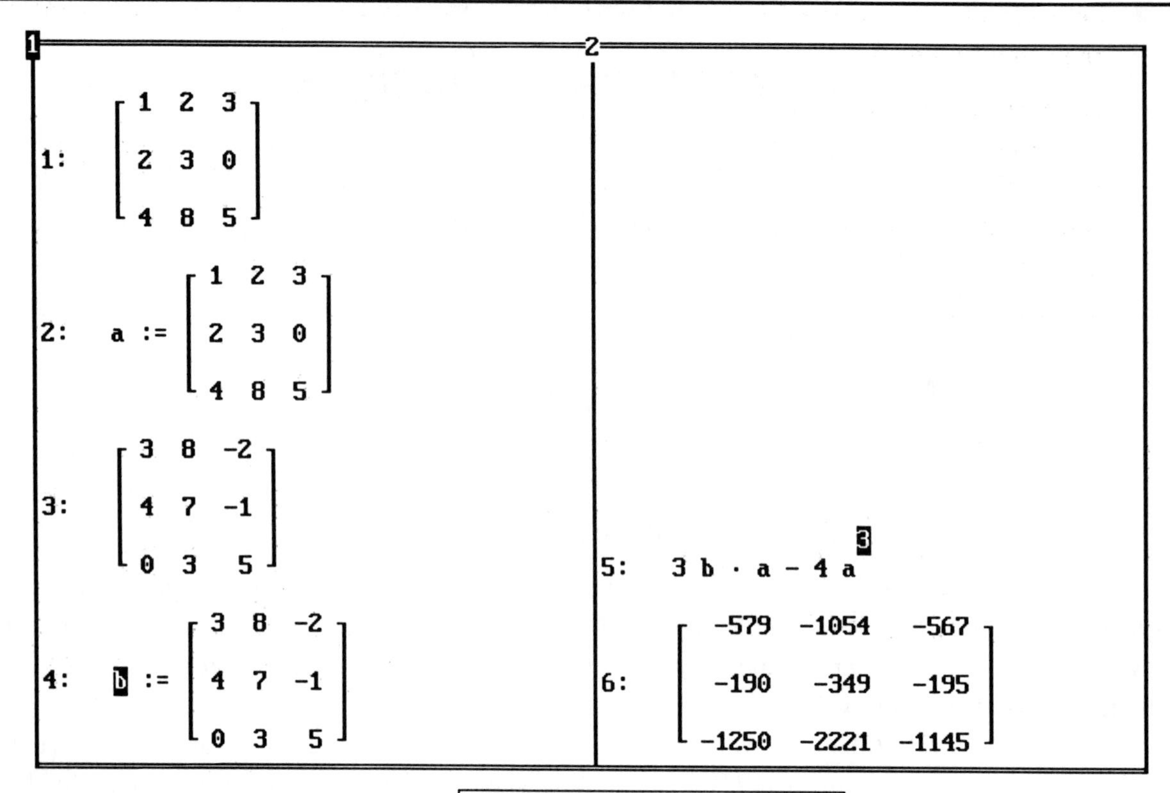

Figure 3.1: Calculating with matrices

Important *DERIVE* Note: The period is essential for matrix products. Neither A∗B nor AB is acceptable. There is another situation to be careful about. If A and B are matrices and you want A^2B, you must **Author** and **Simplify** `A^2 .B` with a space between the 2 and the period. (If you use `A^2.B` with the period next to the 2, *DERIVE* will interpret the period as a decimal point and not as matrix multiplication.) Alternatively you may use parentheses, `(A^2).B`. This is discussed further in item 3 of Section 5 in Appendix I.

Solved Problem 2: Let $A = \begin{bmatrix} 1 & 2 & 3 \\ 4 & 5 & 6 \\ 7 & 8 & 9 \end{bmatrix}$ and $f(x) = x^4 - 3x^2 + 5$. Calculate $f(A)$.

Solution: We must first agree on what $f(A)$ means. If we formally replace x in the definition of f by A (e.g., using the **Manage Substitute** commands), we obtain $A^4 - 3A^2 + 5$, which requires that we add the number 5 to the matrix $A^4 - 3A^2$; but you can't add a number to a matrix. The standard convention in evaluating polynomials at matrices is to replace the constant term by the constant times the identity matrix. Thus, $f(A)$ is defined to be $A^4 - 3A^2 + 5I$, where I is the 3 by 3 identity matrix. *DERIVE*'s syntax for the n by n identity matrix is IDENTITY_MATRIX(n).

Enter the matrix and name it as in Solved Problem 1 above and as seen in Figure 3.2. Next, **Author** and **Simplify** `A^4-3A^2+5identity_matrix(3)`. (You may find it convenient to first make the definition `I:=identity_matrix(3)`. If you do this you can work with the somewhat less cumbersome expression `A^4-3A^2+5I`.) The solution appears as expression 4 of Figure 3.2.

```
1:   [ 1  2  3 ]
     [ 4  5  6 ]
     [ 7  8  9 ]

2:   a := [ 1  2  3 ]
          [ 4  5  6 ]
          [ 7  8  9 ]

      4       2
3:   a  - 3 a  + 5 IDENTITY_MATRIX (3)

4:   [  7475   9180  10890 ]
     [ 16920  20795  24660 ]
     [ 26370  32400  38435 ]
```

Figure 3.2: **A polynomial evaluated at a matrix**

Solved Problem 3: **For students who have studied calculus.** Let $D = \begin{bmatrix} 1 & 0 & 0 \\ 0 & 0 & 0 \\ 0 & 0 & 0.6 \end{bmatrix}$

and $A = \begin{bmatrix} \frac{4}{11} & \frac{17}{11} & -\frac{13}{11} \\ -\frac{14}{11} & \frac{45}{11} & -\frac{26}{11} \\ -\frac{35}{22} & \frac{85}{22} & -\frac{43}{22} \end{bmatrix}$.

Find $\lim_{n\to\infty} D^n$ and $\lim_{n\to\infty} A^n$.

Discussion: An explanation is in order here. Just exactly what does $\lim_{n\to\infty} M^n$ mean when M is a matrix? Definition: $\lim_{n\to\infty} M^n = P$ if $\lim_{n\to\infty} m_{i,j}(n) = p_{i,j}$ where $m_{i,j}(n)$ is the (i,j)th term of M^n. In other words, take the limit of each entry of M^n.

Solution: You should be able to convince yourself that $\begin{bmatrix} a & 0 & 0 \\ 0 & b & 0 \\ 0 & 0 & c \end{bmatrix}^n = \begin{bmatrix} a^n & 0 & 0 \\ 0 & b^n & 0 \\ 0 & 0 & c^n \end{bmatrix}$ for any positive integer n, but you can verify it with *DERIVE*. Therefore,

$$\lim_{n\to\infty} D^n = \lim_{n\to\infty} \begin{bmatrix} 1 & 0 & 0 \\ 0 & 0 & 0 \\ 0 & 0 & 0.6^n \end{bmatrix} = \begin{bmatrix} 1 & 0 & 0 \\ 0 & 0 & 0 \\ 0 & 0 & \lim_{n\to\infty} 0.6^n \end{bmatrix} = \begin{bmatrix} 1 & 0 & 0 \\ 0 & 0 & 0 \\ 0 & 0 & 0 \end{bmatrix}.$$

That was easy. Now, what about A? Unlike the simple diagonal case, there's no apparent simple formula for A^n. We can't just take the nth powers of the elements. So it seems we have no way of actually calculating this limit or even knowing if it exists! But we can look at large powers of A in order to develop some intuition regarding this limit. In Chapter 15 we will see that under the right conditions we can calculate such a limit exactly, but for now we can only estimate it.

```
3:  [a^10, a^20]

4:  [[-0.271485  3.08789  -2.36132]   [-0.272736  3.09093  -2.36362]]
    [[-2.54297   7.17578  -4.72265] , [-2.54547   7.18188  -4.72725]]
    [[-3.17871   7.71973  -4.90332]   [-3.18184   7.72733  -4.90906]]

5:  [a^30, a^40]

6:  [[-0.272740  3.09093  -2.36362]   [-0.272739  3.09094  -2.36362]]
    [[-2.54548   7.18186  -4.72724] , [-2.54548   7.18188  -4.72724]]
    [[-3.18185   7.72732  -4.90906]   [-3.18184   7.72735  -4.90905]]
```

Figure 3.3: **Large powers of a matrix**

Enter the matrix using **Declare Matrix** and name it A using `a:=`F3. We will look at A^{10}, A^{20}, A^{30}, and A^{40}. If we **Author** `[a^10,a^20]` and **approX**, *DERIVE* will approximate both A^{10} and A^{20} as seen in expression 3 of Figure 3.3. In the same way we've approximated A^{30} and A^{40} in expression 5. The corresponding entries in A^{30} and A^{40} are the same to five digits of accuracy, so it seems reasonable that these matrices are good approximations to the desired limit.

Solved Problem 4: If $A = \begin{bmatrix} 2 & 3 & 4 \\ 6 & 4 & 2 \\ 3 & 1 & 6 \end{bmatrix}$, find a polynomial $f(x) = x^3 + px^2 + qx + r$ such that $f(A) = 0$.

Solution: Enter the matrix and name it A. Next **Author** and **Simplify** the expression `A^3+pA^2+qA+r identity_matrix(3)`. *DERIVE* will present the matrix in expression 4 of Figure 3.4. We want each entry in this matrix to be 0. Thus we have a system of 9 equations in 3 unknowns to solve. To do this with *DERIVE* we need to list the nine entries of the matrix as the elements of a vector. The *DERIVE* function APPEND will do this.

DERIVE Note: If you are using an early version of *DERIVE* you will notice that APPEND is not defined. We have provided a definition in Appendix II.

```
                [ 2  3  4 ]
2:   a :=       [ 6  4  2 ]
                [ 3  1  6 ]

      3       2
3:   a  + p a  + q a + r IDENTITY_MATRIX (3)

     [ 34 p + 2 q + r + 314    22 p + 3 q + 228       38 p + 4 q + 408     ]
4:   [ 42 p + 6 q + 432        36 p + 4 q + r + 314   44 p + 2 q + 504     ]
     [ 30 p + 3 q + 324        19 p + q + 216         50 p + 6 q + r + 458 ]

            [ 34 p + 2 q + r + 314    22 p + 3 q + 228       38 p + 4 q + 408     ]
5:   APPEND [ 42 p + 6 q + 432        36 p + 4 q + r + 314   44 p + 2 q + 504     ]
            [ 30 p + 3 q + 324        19 p + q + 216         50 p + 6 q + r + 458 ]

6:   [34 p + 2 q + r + 314, 22 p + 3 q + 228, 38 p + 4 q + 408, 42 p + 6 q + 432

7:   [p = -12, q = 12, r = 70]
```

Figure 3.4: A polynomial that annihilates a matrix

Author and **Simplify** `append(#4)` to obtain the list of equations in expression 6 of Figure 3.4. Ask *DERIVE* to **soLve**, and the solution can be read from expression 7 as $f(x) = x^3 - 12x^2 + 12x + 70$. You should check this answer to ensure that $f(A) = 0$. Solved Problem 2 explains how to evaluate a polynomial at a matrix.

Note: There is something quite remarkable about this result. We solved a system of nine equations in three unknowns and got a unique solution! This says that there is one and only one cubic polynomial f (with leading coefficient 1) for which $f(A) = 0$. We'll return to this topic in a later chapter. See the exercises for more.

Exercises

1. If $A = \begin{bmatrix} 1 & 2 & 3 & 0 \\ -1 & 1 & 0 & -2 \\ 2 & 3 & 1 & 0 \\ 4 & 2 & 6 & -3 \end{bmatrix}$ and $B = \begin{bmatrix} 3 & -1 & 2 & 4 \\ 0 & 1 & 2 & 3 \\ -4 & 2 & 0 & -1 \\ 2 & 1 & 4 & 3 \end{bmatrix}$, calculate the following. Comment on how the result compares to the case when A and B are replaced by numbers. (See Exploration and Discovery Problem 1.)

   (a) $A^2B^2 - (AB)^2$

   (b) $ABA - BA^2$

   (c) $A^2 + 2A + I - (A + I)^2$

2. Evaluate the following polynomials at A, where A is as in Exercise 1.

   (a) $x^2 - 3$

   (b) $(x^2 - 3)^2$

   (c) $x^4 - 6x^2 + 9$

   (d) Discuss the relation between results in (b) and (c).

3. **For students who have studied calculus.** Approximate $\lim_{n\to\infty} A^n$ in each case and discuss your reason for choosing the value of n you decided to use for the approximation.

   (a) $A = \begin{bmatrix} \frac{21}{4} & -\frac{7}{2} & \frac{13}{8} \\ \frac{21}{4} & -\frac{7}{2} & \frac{17}{8} \\ -\frac{9}{2} & 3 & \frac{1}{4} \end{bmatrix}$

   (b) $A = \begin{bmatrix} 23 & -16 & 7 \\ 33 & -23 & \frac{21}{2} \\ 0 & 0 & 1 \end{bmatrix}$

(c) $A = \begin{bmatrix} \frac{1}{7} & \frac{2}{7} & \frac{3}{7} \\ \frac{4}{7} & \frac{3}{7} & \frac{5}{7} \\ \frac{6}{7} & 0 & \frac{2}{7} \end{bmatrix}$

4. **For students who have studied calculus.** Estimate $\lim_{n\to\infty}\left(I+\frac{A}{n}\right)^n$ where A is $\begin{bmatrix} -7 & 46 & -21 \\ -12 & 79 & -36 \\ -24 & 158 & -72 \end{bmatrix}$, and I is the 3 by 3 identity matrix. Discuss your reason for choosing the value of n you decided to use for the approximation. If you've had Calculus, you should know what $\lim_{n\to\infty}\left(1+\frac{x}{n}\right)^n$ is if x is a real number. What is it?

5. Let A be the matrix in Solved Problem 4.

   (a) Show there is no quadratic polynomial $g(x) = x^2 + px + q$ such that $g(A) = 0$.

   (b) Find a polynomial of the form $f(x) = x^4 + px^3 + qx^2 + rx + s$ such that $f(A) = 0$.

   (c) Is there a unique answer to (b)? If not, describe them all.

   (d) Carefully examine your answer to (c) and compare it to the cubic polynomial we found in Solved Problem 4. Explain what you notice.

6. Describe all 2 by 2 matrices that commute with $A = \begin{bmatrix} 2 & -1 \\ 3 & 5 \end{bmatrix}$.

   (Hint. Let $B = \begin{bmatrix} p & q \\ r & s \end{bmatrix}$ and consider $AB - BA$=0.)

7. Describe all 3 by 3 matrices that commute with $A = \begin{bmatrix} 1 & 2 & -1 \\ 3 & 0 & 1 \\ 5 & -1 & 1 \end{bmatrix}$.

   (Hint. Let $B = \begin{bmatrix} p & q & r \\ s & t & u \\ v & w & x \end{bmatrix}$ and consider $AB - BA$=0.)

8. Describe all 3 by 3 matrices that commute with $\begin{bmatrix} -7 & 3 & -3 \\ 30 & 2 & 6 \\ 45 & -9 & 17 \end{bmatrix}$.

9. **Markov chains**: Each year a certain number of people move from city A to city B, while others move from city B to city A. Statistically it is found that the probability of a citizen of city A moving to city B in a given year is 0.04, while the probability of a citizen of city B moving to city A is 0.03. For purposes of this exercise we assume that no other factors

influence the populations. Thus, if a_n represents the number of people in city A in year n and b_n represents the number of people in city B, then in year $n+1$ we expect to find $0.96a_n + 0.03b_n$ people in city A and $0.04a_n + 0.97b_n$ people in city B.

This says that if $A = \begin{bmatrix} 0.96 & 0.03 \\ 0.04 & 0.97 \end{bmatrix}$, then $A \begin{bmatrix} a_n \\ b_n \end{bmatrix} = \begin{bmatrix} a_{n+1} \\ b_{n+1} \end{bmatrix}$.

Processes of this sort in which the state at time $n+1$ depends only on the state at time n are known as *Markov chains.* A is known as the *transition matrix.* (See *Elementary Linear Algebra, Applications Version* by Howard Anton and Chris Rorres for a precise definition as well as a deeper analysis of the topic.)

(a) Show that in general $A^n \begin{bmatrix} a_0 \\ b_0 \end{bmatrix} = \begin{bmatrix} a_n \\ b_n \end{bmatrix}$.

In the three four parts, assume that in 1990 the population of city A was 240,000 and the population of city B was 196,000.

(b) What population figures would you predict for the year 1995?

(c) What do you expect the population to be in the years 2000 and 2010?

(d) What population figures do you expect the cities will reach far into the future?

10. **More on Markov chains:** Probabilities for population movements among three cities are given in the table below.

	From A	From B	From C
Probability of moving to A	0.7	0.1	0.05
Probability of moving to B	0.2	0.8	0.05
Probability of moving to C	0.1	0.1	0.90

The initial population of city A is 375,000. Find the populations of cities B and C if the total population of the three cities remains constant from year to year.

Exploration and Discovery

1. The following is a list of familiar formulas that are true for real numbers. Decide which are true for matrices by testing each of them using 2 by 2 matrices with variable entries. Provide specific numerical counterexamples for those that are false and try to prove those that are true. (Compare the first three of them to Exercise 1.)

   (a) $AB = BA$

   (b) $A^2B^2 = (AB)^2$

   (c) $A^2 + 2A + I = (A + I)^2$

   (d) $(A + B)^2 = A^2 + 2AB + B^2$

   (e) $(A + B)(A - B) = A^2 - B^2$

   (f) $A^2A^3 = A^3A^2$

   (g) Test other familiar formulas.

2. **This exercise uses the file REDUCE.MTH. See Appendix II.** Let A be an n by k matrix, let I denote the n by n identity matrix, and let B denote the augmented matrix $\left[\begin{array}{c|c} A & I \end{array}\right]$. The command REDUCE(B,$k$) will return an augmented matrix $\left[\begin{array}{c|c} C & D \end{array}\right]$ where C is the reduced echelon form of A. We wish to investigate the nature of the matrix D. For example, if $A = \begin{bmatrix} 1 & 2 & 3 \\ 4 & 5 & 6 \\ 7 & 8 & 9 \end{bmatrix}$ then $D = \begin{bmatrix} 0 & -\frac{8}{3} & \frac{5}{3} \\ 0 & \frac{7}{3} & -\frac{4}{3} \\ 1 & -2 & 1 \end{bmatrix}$. (To check this result enter the matrix $\begin{bmatrix} 1 & 2 & 3 & 1 & 0 & 0 \\ 4 & 5 & 6 & 0 & 1 & 0 \\ 7 & 8 & 9 & 0 & 0 & 1 \end{bmatrix}$ and name it B. Next **Author** and **Simplify** `reduce(B,3)`.) Calculate the product DA for this example.

   For several other choices of A (including matrices that are not square) calculate the product DA. Based on your work, formulate a conjecture about the matrix D and the reduced echelon form of A.

3. **For students who have studied calculus.** We've discussed what it means to evaluate a polynomial at a matrix, but what does cos(A) mean if A is a matrix? (You can ask *DERIVE* but you won't get an answer.) Here's an idea: $\cos x$ has a Maclaurin series expansion $1 + \frac{x^2}{2} + \frac{x^4}{24} + \cdots$ whose partial sums are polynomials which *can* be evaluated at a matrix. Perhaps these can provide an approximation to "$\cos A$." As an example, we will approximate $\cos A$, where A is $\begin{bmatrix} -7 & 46 & -21 \\ -12 & 79 & -36 \\ -24 & 158 & -72 \end{bmatrix}$.

   Enter the matrix and name it A as in expression 2 of Figure 3.5. Now **Author** **`taylor(cos x, x, 0, 6) -1 + identity_matrix(3)`** as seen in expression 3 of Figure 3.5 and **Simplify**. (The reason for this is to replace the "1" in the polynomial by the 3 by 3 identity matrix.) Now use **Manage Substitute** to replace x by A as seen in expression 5 of Figure 3.5. Finally, **approX** expression 5 *(***approX***imate, do not simplify)* to obtain the approximation in expression 6 of Figure 3.5.

2: $a := \begin{bmatrix} -7 & 46 & -21 \\ -12 & 79 & -36 \\ -24 & 158 & -72 \end{bmatrix}$

3: `TAYLOR (COS (x), x, 0, 6) - 1 + IDENTITY_MATRIX (3)`

4: $\begin{bmatrix} 1 & 0 & 0 \\ 0 & 1 & 0 \\ 0 & 0 & 1 \end{bmatrix} - \frac{x^6}{720} + \frac{x^4}{24} - \frac{x^2}{2}$

5: $\begin{bmatrix} 1 & 0 & 0 \\ 0 & 1 & 0 \\ 0 & 0 & 1 \end{bmatrix} - \frac{a^6}{720} + \frac{a^4}{24} - \frac{a^2}{2}$

6: $\begin{bmatrix} 0.540277 & 2.75833 & -1.37916 \\ 0 & 0.540277 & 0 \\ 0 & -0.919444 & 1 \end{bmatrix}$

Figure 3.5: **Approximating the cosine of a matrix**

(a) Repeat the process above using the Maclaurin polynomial of degrees 10 and 20. (Note: Depending on your machine, it may take several seconds for *DERIVE* to do this.) How do the three compare?

(b) Repeat the process above to approximate $\sin A$ using Maclaurin polynomials of degrees 6, 10, and 20. (Explain why you don't see an x^6 term in the polynomial of degree 6.)

(c) Square the matrices above that you used to approximate $\cos A$ and $\sin A$ and add them. Does the formula $\sin^2 x + \cos^2 x = 1$ (or one like it) appear to hold for matrices? Explain.

(d) Repeat the process above to approximate e^A using Maclaurin polynomials of degrees 6, 10 and 20. Compare your answers to the one you obtained in exercise 4.

(e) Does the formula $e^{2x} = (e^x)^2$ appear to hold when x is a matrix?

4. (a) Do exercise 6.

(b) If A is the matrix in exercise 6, use *DERIVE* to find $f(A)$ where $f(x) = px^2 + qx + r$. (Use variable coefficients as shown.) Using your answer to this and part (a) show that every matrix that commutes with A is a quadratic polynomial in A.

Remark: It's easy to see that if $f(x)$ is a polynomial and A is a matrix then $f(A)$ commutes with A. The remarkable thing is that the converse is true in this case.

(c) Do Exercise 7.

(d) Let A be the matrix from Exercise 7. Is it true that every matrix that commutes with A is a quadratic polynomial in A? If it is true, prove it. If not, give a counter-example and prove your example is not a quadratic polynomial in A.

5. **For students who have studied calculus.** The *DERIVE* commands **Calculus Differentiate** produce the derivative of the highlighted function. If you ask *DERIVE* to differentiate a matrix, it will differentiate each entry of the matrix. For example, enter the matrix $\begin{bmatrix} x & x^2 \\ x^3 & \sin x \end{bmatrix}$. With it highlighted, execute the **Calculus Differentiate** commands, accepting the defaults by pressing Enter. *DERIVE* will present $\frac{d}{dx}\begin{bmatrix} x & x^2 \\ x^3 & \sin x \end{bmatrix}$.

Simplify this and you get $\begin{bmatrix} 1 & 2x \\ 3x^2 & \cos x \end{bmatrix}$.

Make up two 3 by 3 matrices whose entries are functions of x and test the formula $\frac{d}{dx}AB = (\frac{d}{dx}A)B + A(\frac{d}{dx}B)$. Does this usual product rule for derivatives appear to hold for matrices? Can you prove it? Try similar tests for other differentiation formulas such as the chain rule. Discuss your findings.

LABORATORY EXERCISE 3.1

Polynomials and Matrices

Name ______________________________ **Due Date** ____________________

The problem: In algebra you are given problems such as this: Show that $f(1) = 0$ if $f(x) = x^2 + 3x - 4$. As we have seen, we can ask about $f(A)$ when A is a matrix. In Solved Problem 4 and Exercise 5, we examined some questions about finding polynomials f such that $f(A) = 0$. (In this context, 0 means the zero *matrix,* of course.) In this problem we want to examine this idea. In each part of this lab exercise, be sure to discuss how you arrived at your answer. You may also want to attach printouts of *DERIVE* screens to refer to in your explanations. Let

$$A = \begin{bmatrix} 1 & 2 & 3 & 0 \\ -1 & 1 & 0 & -2 \\ 2 & 3 & 1 & 0 \\ 4 & 2 & 6 & -3 \end{bmatrix}$$

1. Is there a quadratic polynomial $h(x) = x^2 + px + q$ such that $h(A) = 0$?

2. Is there a cubic polynomial $g(x) = x^3 + px^2 + qx + r$ such that $g(A) = 0$?

continued on next page

3. Find a polynomial $f(x) = x^4 + px^3 + qx^2 + rx + s$ such that $f(A) = 0$.

4. In the cases above where such a polynomial exists, is it unique? If not, describe all polynomials f for which $f(A) = 0$.

5. Based on Solved Problem 4 and your work here, venture a guess about how the size of a square matrix M and the degree of a polynomial $p(x)$ are related when $p(M) = 0$.

6. Let $B = \begin{bmatrix} 2 & 0 \\ 0 & 2 \end{bmatrix}$ and $k(x) = x - 2$. Find $k(B)$. Does this suggest a modification of your conjecture?

LABORATORY EXERCISE 3.2

A Genetically Transmitted Plant Disease

Name ______________________________ **Due Date** ____________________

The problem: A field of plants is infected with a genetically transmitted disease. After n years there are x_n diseased plants, y_n plants that are carriers of the disease but do not show any symptoms of the disease, and z_n disease-free plants. (Initially there are x_0 diseased plants, y_0 carriers, and z_0 healthy plants.) We assume that each plant produces a single offspring each year. Half the offspring of the diseased plants die, and the remaining half are diseased. The disease-free plants produce disease-free offspring. The carriers produce $\frac{1}{3}$ diseased plants, $\frac{1}{3}$ carriers, and $\frac{1}{3}$ healthy plants. The goal of this exercise is to trace the course of the disease.

1. Find a formula for each of x_{n+1}, y_{n+1}, and z_{n+1} in terms of x_n, y_n, and z_n.

2. Find a matrix A such that $A \begin{bmatrix} x_n \\ y_n \\ z_n \end{bmatrix} = \begin{bmatrix} x_{n+1} \\ y_{n+1} \\ z_{n+1} \end{bmatrix}$.

continued on next page

3. Use a matrix equation to show how to get $\begin{bmatrix} x_n \\ y_n \\ z_n \end{bmatrix}$ from $\begin{bmatrix} x_0 \\ y_0 \\ z_0 \end{bmatrix}$.

4. Describe the field after 10 years. That is, how many plants are diseased, how many are carriers, and how many are disease-free?

5. Describe the field after 20 years.

6. From the data in parts 4 and 5 above, what will the field *eventually* look like? That is, after many years, what can you say about the numbers of diseased plants, carriers, and healthy plants?

CHAPTER 4

INVERSES OF MATRICES

LINEAR ALGEBRA CONCEPTS

- **Matrix inversion**

Introduction

If a is a non-zero number, its multiplicative inverse, or reciprocal, is the number b such that $ab = 1$. Similarly, if A is a square matrix, its inverse, or reciprocal, is a matrix B such that $AB = BA = I$, if such a matrix exists. As we will see, this idea is important for matrices just as it is for numbers.

Solved Problems

Solved Problem 1: Find the inverses of $A = \begin{bmatrix} 1 & 2 & 4 \\ 3 & 5 & 9 \\ 2 & 6 & 7 \end{bmatrix}$ and $B = \begin{bmatrix} 1 & 2 & 3 \\ 4 & 5 & 6 \\ 7 & 8 & 9 \end{bmatrix}$.

Solution: Enter the two matrices using **Declare Matrix** and name them A and B as seen in Figure 4.1. **Author** and **Simplify** `a^-1`. The inverse is expression 3 of Figure 4.1. If we do the same for B, *DERIVE* will simply return expression 6 indicating that no inverse exists.

Alternative Solution: We can use *DERIVE* to emulate the familiar calculation of inverses by row-reducing the augmented matrices $\left[\begin{array}{c|c} A & I \end{array}\right]$ and $\left[\begin{array}{c|c} B & I \end{array}\right]$. *DERIVE*'s ROW_REDUCE command has the facility to build and then row reduce the augmented matrices in a single step. Recall that IDENTITY_MATRIX(3) is *DERIVE*'s name for the 3 by 3 identity matrix.

To get the inverse of A **Author** and **Simplify** `row_reduce(A,identity_matrix(3))`. From expression 9 of Figure 4.2 we see that A reduces to the identity matrix. We conclude that A has an inverse, and it appears as the last three columns of expression 9. If we perform the same procedure for B, we see from expression 11 of Figure 4.2 that the reduced form of B is not the identity matrix. Hence B has no inverse.

1: $a := \begin{bmatrix} 1 & 2 & 4 \\ 3 & 5 & 9 \\ 2 & 6 & 7 \end{bmatrix}$

2: a^{-1}

3: $\begin{bmatrix} -\frac{19}{7} & \frac{10}{7} & -\frac{2}{7} \\ -\frac{3}{7} & -\frac{1}{7} & \frac{3}{7} \\ \frac{8}{7} & -\frac{2}{7} & -\frac{1}{7} \end{bmatrix}$

4: $b := \begin{bmatrix} 1 & 2 & 3 \\ 4 & 5 & 6 \\ 7 & 8 & 9 \end{bmatrix}$

5: b^{-1}

6: $\begin{bmatrix} 1 & 2 & 3 \\ 4 & 5 & 6 \\ 7 & 8 & 9 \end{bmatrix}^{-1}$

Figure 4.1: Direct method for finding inverses

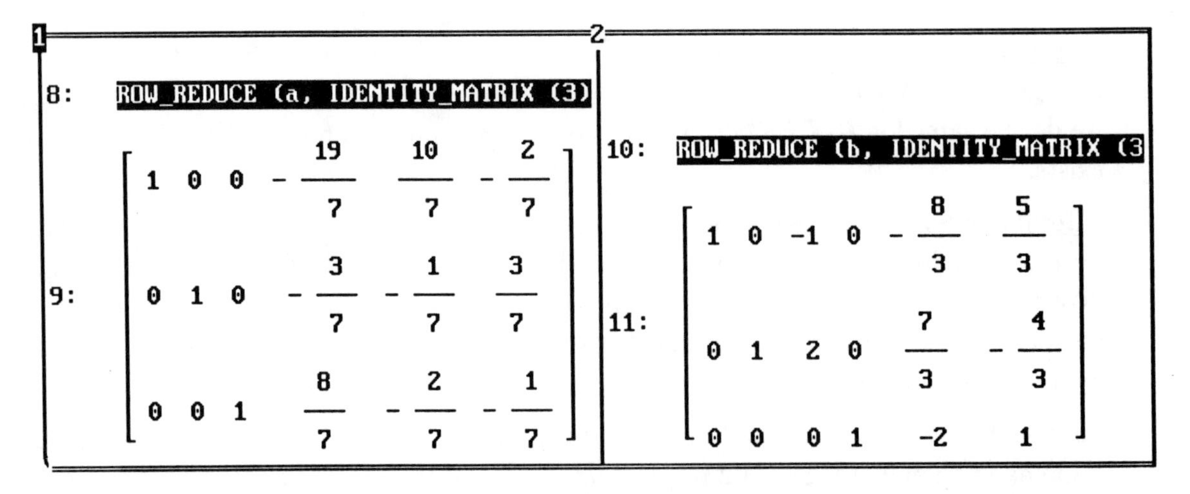

8: ROW_REDUCE (a, IDENTITY_MATRIX (3))
9: $\begin{bmatrix} 1 & 0 & 0 & -\frac{19}{7} & \frac{10}{7} & -\frac{2}{7} \\ 0 & 1 & 0 & -\frac{3}{7} & -\frac{1}{7} & \frac{3}{7} \\ 0 & 0 & 1 & \frac{8}{7} & -\frac{2}{7} & -\frac{1}{7} \end{bmatrix}$
10: ROW_REDUCE (b, IDENTITY_MATRIX (3
11: $\begin{bmatrix} 1 & 0 & -1 & 0 & -\frac{8}{3} & \frac{5}{3} \\ 0 & 1 & 2 & 0 & \frac{7}{3} & -\frac{4}{3} \\ 0 & 0 & 0 & 1 & -2 & 1 \end{bmatrix}$

Figure 4.2: The inverse from the reduced echelon form of an augmented matrix

Solved Problem 2: Let A and B denote the matrices in Solved Problem 1. Find a 3 by 3 matrix C that satisfies the following equation.

$$A^2(4C - 3B)A^{-1} = BA - AB$$

Solution: The first step is to solve the equation by hand. The steps are shown below. Notice that when we multiply the equation by a matrix we use the same order of multiplication on each side.

$$\begin{aligned}
A^2(4C - 3B)A^{-1} &= BA - AB \\
A^2(4C - 3B) &= (BA - AB)A \\
4C - 3B &= A^{-2}(BA - AB)A \\
4C &= A^{-2}(BA - AB)A + 3B \\
C &= \frac{1}{4}(A^{-2}(BA - AB)A + 3B)
\end{aligned}$$

5: $\frac{1}{4}(a^{-2} \cdot (b \cdot a - a \cdot b) \cdot a + 3\, b)$

6: $\begin{bmatrix} \frac{3119}{28} & \frac{1782}{7} & \frac{10253}{28} \\ \frac{55}{28} & \frac{23}{28} & \frac{15}{28} \\ -\frac{773}{28} & -\frac{961}{14} & -\frac{2827}{28} \end{bmatrix}$

Figure 4.3: **Solving an equation for a matrix**

The matrices A and B should be entered and defined as in Solved Problem 1. **Author** `1/4(A^-2 .(B.A-A.B).A+3B)`. (Note: Don't forget the periods to indicate matrix multiplication, and recall that the space between −2 and the period is necessary. Refer to item 3(a) in Section 5 of Appendix I to refresh your memory if necessary.) If we **Simplify** expression 5 in Figure 4.3, we get the answer in expression 6.

Solved Problem 3: Use the inverse to solve the following system of equations.

$$\begin{aligned} 7x + 4y - 2z + 4w &= 8 \\ 2x - 3y + 7z - 6w &= 4 \\ 5x + 6y + 2z - 5w &= 2 \\ 3x + 3y - 5z + 8w &= 9 \end{aligned}$$

Solution: If $A = \begin{bmatrix} 7 & 4 & -2 & 4 \\ 2 & -3 & 7 & -6 \\ 5 & 6 & 2 & -5 \\ 3 & 3 & -5 & 8 \end{bmatrix}$, $\mathbf{u} = \begin{bmatrix} x \\ y \\ z \\ w \end{bmatrix}$ and $\mathbf{v} = \begin{bmatrix} 8 \\ 4 \\ 2 \\ 9 \end{bmatrix}$, then the system of equations is equivalent to the matrix equation $A\mathbf{u} = \mathbf{v}$. Multiplying each side of the equation on the left by A^{-1} produces the solution $\mathbf{u} = A^{-1}\mathbf{v}$. After the matrix A is entered and named, **Author** `A^-1 .[8,4,2,9]`. (Note: Notice we did *not* author `[8,4,2,9]'` as you might see in your text. See Section 4 of Appendix I for an explanation. Also watch the space after the -1.) The solution $x = -\frac{29}{28}$, $y = \frac{21}{8}$, $z = \frac{295}{56}$, $w = \frac{107}{28}$ can now be read from expression 4 of Figure 4.4.

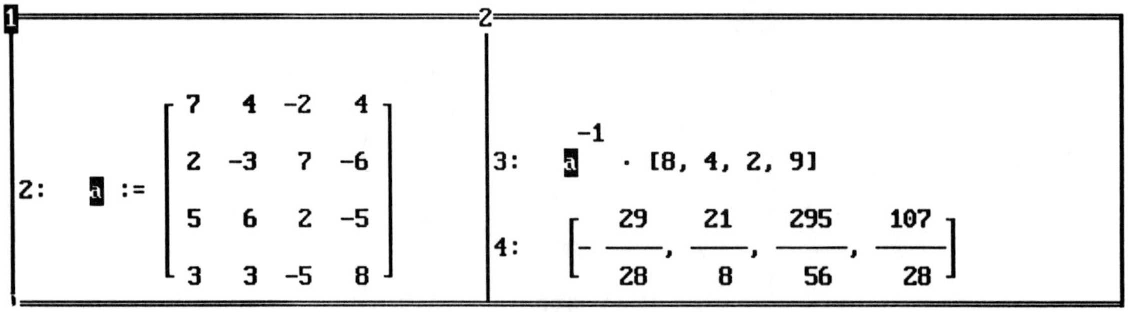

```
1                                  2
2:  a :=  [ 7   4  -2   4 ]   |  3:  a^-1 · [8, 4, 2, 9]
          [ 2  -3   7  -6 ]   |
          [ 5   6   2  -5 ]   |  4:  [- 29/28, 21/8, 295/56, 107/28]
          [ 3   3  -5   8 ]   |
```

Figure 4.4: Using the inverse to solve a system of equations

Exercises

1. Use matrix inverses to solve the following systems of equations.

(a)
$$\begin{aligned} 7x + 4y - 2z &= 8 \\ 4x + 7y + 5z &= 5 \\ 2x - 3y + 8z &= 2 \end{aligned}$$

(b)
$$\begin{aligned} 2x - 5y + z &= a \\ 7x - y + 4z &= b \\ 3x - 6y + 5x &= 9 \end{aligned}$$

(c)
$$\begin{aligned} ax + 4y + 2z &= 8 \\ 3x + by - 5z &= 1 \\ 4x + 3y + cz &= 3 \end{aligned}$$

2. For what values of s and t does $\begin{bmatrix} 2 & 7 & t \\ 3 & s & 1 \\ 2 & 4 & 6 \end{bmatrix}$ have an inverse?

3. Suppose that neither $\begin{bmatrix} 1 & 2 & s \\ 2 & 3 & t \\ 4 & 5 & 7 \end{bmatrix}$ nor $\begin{bmatrix} 4 & 5 & 8 \\ s & 2 & 3 \\ t & 1 & 8 \end{bmatrix}$ has an inverse. Find s and t.

4. Let $C = \begin{bmatrix} 3 & 4 & 1 \\ 9 & 2 & 7 \\ 4 & 1 & 5 \end{bmatrix}$ and $D = \begin{bmatrix} 5 & 6 & 2 \\ 9 & 4 & 7 \\ 2 & 2 & 1 \end{bmatrix}$. Solve the following equations for the 3 by 3 matrix A.

   (a) $CAD = C - D$

(b) $C^2(A+2D) = DA$

(c) $3AD^{-1} = A^2DC$ (Assume that A is invertible.)

5. If A is a square matrix and $f(x) = \frac{p(x)}{q(x)}$, where p and q are polynomials, we define $f(A)$ to be $p(A)q(A)^{-1}$ provided that $q(A)^{-1}$ exists. (Recall that to evaluate a polynomial at a matrix the constant term must be multiplied by the identity matrix. See also Exploration and Discovery Problem 3.)

(a) If $A = \begin{bmatrix} 5 & 6 \\ 2 & 8 \end{bmatrix}$, evaluate the following rational functions at A.

(i) $\frac{x^3+2}{x^2+1}$ (ii) $\frac{x^2+x-2}{x-1}$

(b) Show that $\frac{x^2+x-2}{x-1} = x+2$ if $x \neq 1$. Evaluate both sides for $x = A$. Are they the same?

(c) If $B = \begin{bmatrix} 29 & 45 \\ -18 & -28 \end{bmatrix}$, evaluate $\frac{x^2+x-2}{x-1}$ and $x+2$ at B. Are they the same? Explain what you observe.

6. Let $A = \begin{bmatrix} 1 & 3 & 2 \\ 4 & 5 & 1 \\ 3 & 7 & 2 \end{bmatrix}$. Find a quadratic polynomial $f(x)$ such that $f(A) = A^{-1}$.
(Hint: Find p, q, and r such that $pA^2 + qA + rI = A^{-1}$, where I is the 3 by 3 identity matrix. Remember that APPEND will change a matrix of equations to a vector of equations that *DERIVE* can **soLve**.)

7. In Solved Problem 4 of Chapter 3, we found a cubic polynomial $f(x)$ such that $f(A) = 0$. Show that the matrix A is invertible. Use the solution to Solved Problem 4 of Chapter 3 to express A^{-1} as a quadratic polynomial in A.

Prove that if a matrix is invertible and satisfies a polynomial equation $f(A) = 0$, then A^{-1} can be expressed as a polynomial in A. (No computers here.)

8. **Markov chains revisited**: This problem refers to Exercise 9 in Chapter 3. By looking at powers of the transition matrix in that exercise we could predict population trends. Powers of the inverse of the matrix tell us what happened in the past.

What would you expect to find if you looked up the population figures for the cities in 1985? In 1980?

Exploration and Discovery

1. Try to decide which of the following equations are true for matrices. For those that are, give a proof. For those that are not, give a counterexample. Use *DERIVE* to verify your counterexamples.

   (a) $(AB)^{-1} = A^{-1}B^{-1}$

   (b) $(A^3)^{-1} = (A^{-1})^3$

   (c) $(A^{-1}BA)^3 = A^{-1}B^3A$

   (d) $(ABC)^{-1} = C^{-1}B^{-1}A^{-1}$

   (e) Make up other formulas that are true for numbers but not matrices.

2. An upper triangular matrix is a matrix all of whose entries below the main diagonal are zero. For example, $\begin{bmatrix} 1 & 2 & 3 & 4 \\ 0 & 3 & 2 & 3 \\ 0 & 0 & 1 & 2 \\ 0 & 0 & 0 & 5 \end{bmatrix}$ is upper triangular. Calculate the inverses of several 4 by 4 upper triangular matrices. Is the inverse of an upper triangular matrix necessarily upper triangular? Prove your answer.

3. First do Exercise 5 where we defined $f(A)$ to be $p(A)q(A)^{-1}$.

   (a) Repeat the exercise using the definition $f(A) = q(A)^{-1}p(A)$. Compare your answers and discuss the results. Do you suspect $p(A)q(A)^{-1} = q(A)^{-1}p(A)$ for any square matrix A for which $q(A)$ is invertible? Can you prove it?

   (b) Now consider polynomials in two variables such as $p(x,y) = xy^2+3$ or $q(x,y) = 2x+y$. Calculate $p(A,B)$ and $q(A,B)$ when A and B are the matrices in Solved Problem 1.

   (c) Explore the two methods of defining $f(A,B)$ where $f(x,y) = \dfrac{p(x,y)}{q(x,y)}$ and discuss your findings.

4. **For students who have studied calculus.** In Exploration and Discovery Problem 3 in Chapter 3, we discussed how one might define $\cos(A)$ and e^A, where A is a square matrix, by using Maclaurin series. Since $\dfrac{1}{\cos(x)} = \sec(x)$ and $\dfrac{1}{e^x} = e^{-x}$, we may be led to conjecture that the inverses of the matrices $\cos(A)$ and e^A are, respectively, $\sec(A)$ and e^{-A}. (How would you define $\sec(A)$ for a matrix A?) By working with the series, try to find evidence that the conjectures are true. Discuss your observations and findings.

LABORATORY EXERCISE 4.1

ANIMAL MIGRATION

Name ______________________________ **Due Date** ________________

Before beginning, it may be helpful to look at Exercise 8.

The problem: The state of Maine has been subdivided into four regions labeled A, B, C, and D to study migration patterns of a certain species of animal. In a given year some animals move from one region to another while some remain where they are. Studies have shown that animals from a given region move to other regions with probabilities given by the following table.

	From A	From B	From C	From D
Probability of moving to A	0.7	0.05	0.1	0.2
Probability of moving to B	0.1	0.60	0.2	0.1
Probability of moving to C	0.1	0.20	0.5	0.1
Probability of moving to D	0.1	0.15	0.2	0.6

In 1990 the populations of the regions were

A: 5500 B: 4800 C: 4100 D: 5250

For purposes of this exercise assume that these are the only factors governing the animal populations in these regions.

1. Let a_n denote the population of region A in year n, b_n the population of B in year n, c_n the population of C in year n and d_n the population of D in year n. Let $\mathbf{p}_n$ denote the vector $\begin{bmatrix} a_n \\ b_n \\ c_n \\ d_n \end{bmatrix}$. If $A = \begin{bmatrix} 0.7 & 0.05 & 0.1 & 0.2 \\ 0.1 & 0.6 & 0.2 & 0.1 \\ 0.1 & 0.2 & 0.5 & 0.1 \\ 0.1 & 0.15 & 0.2 & 0.6 \end{bmatrix}$, show that $A\mathbf{p}_n = \mathbf{p}_{n+1}$.

continued on next page

2. Use a matrix equation to show how to get $\mathbf{p}_n$ from $\mathbf{p}_0$.

3. What are the expected populations of the four regions in 1995 and 2000?

4. Describe the eventual populations of the four regions. Explain how you arrived at your answer.

5. If you examined the regions in 1985, what population would you expect to find?

6. A prominent ecologist contends that this migration pattern has been persisting for at least 10 years. Criticize this statement.

CHAPTER 5

DETERMINANTS, ADJOINTS, AND CRAMER'S RULE

LINEAR ALGEBRA CONCEPTS

- **Determinant**
- **Adjoint**
- **Cramer's rule**

Introduction

The determinant is a function that assigns to each square matrix A a real number $\text{DET}(A)$. This number is important in determining if a matrix has an inverse and can be used to directly obtain solutions of certain systems of equations.

Some books use $|A|$ for the determinant of the matrix A and some use $\text{DET}(A)$. *DERIVE* uses the latter syntax. (Exercise 1 asks you to compare the answers *DERIVE* gives for the two.)

Solved Problems

Solved Problem 1: **This exercise uses the file VECTOR.MTH..** Calculate the determinant and the adjoint of $\begin{bmatrix} 4 & 2 & -3 & 5 \\ 7 & 3 & 9 & 8 \\ 4 & 5 & 3 & 2 \\ 6 & 6 & 5 & 4 \end{bmatrix}$.

Solution: Enter the matrix using **Declare Matrix**. Now **Author** `det(#1)` and **Simplify**. The answer is expression 3 of Figure 5.1.

To obtain the adjoint, we must load the VECTOR.MTH file using **Transfer Merge** VECTOR. After it is loaded, **Author** and **Simplify** `adjoint(#1)`. The adjoint appears in expression 23 of Figure 5.1.

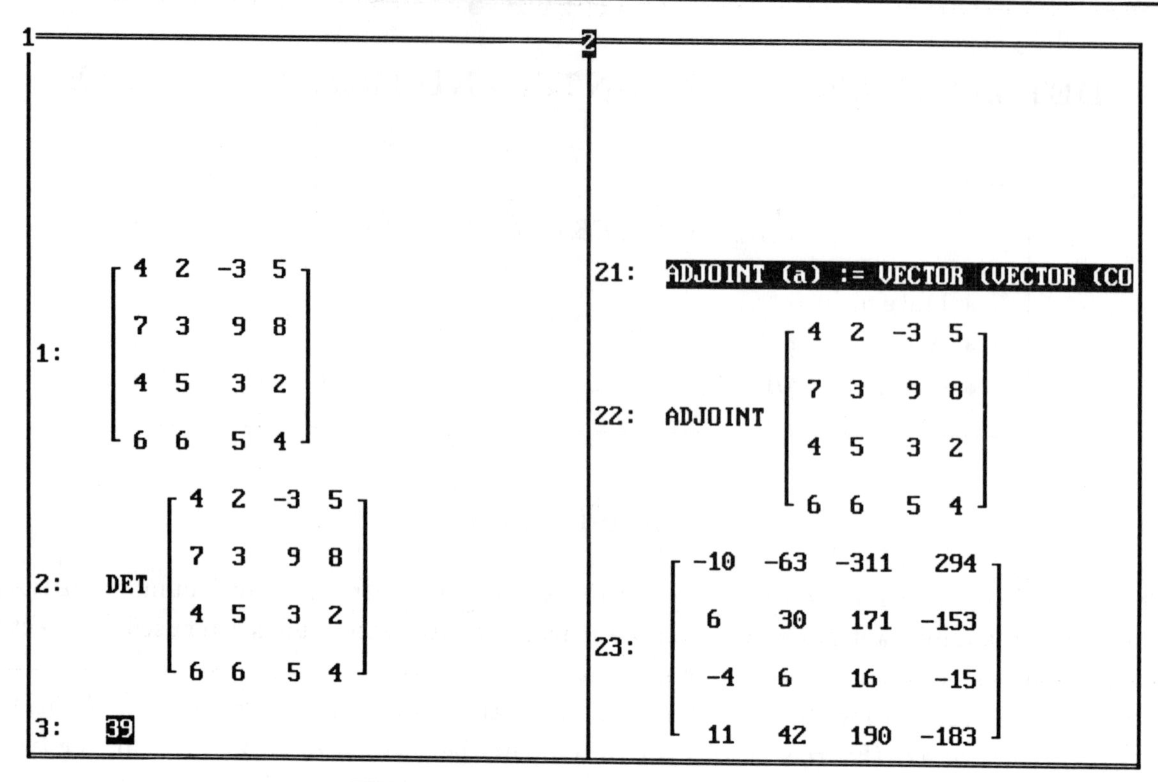

Figure 5.1: The determinant and adjoint of a matrix

Solved Problem 2: For the matrix in Solved Problem 1, verify that if any multiple of row 2 is added to row 1, the value of the determinant does not change.

Solution: Enter the matrix $\begin{bmatrix} 4+7t & 2+3t & -3+9t & 5+8t \\ 7 & 3 & 9 & 8 \\ 4 & 5 & 3 & 2 \\ 6 & 6 & 5 & 4 \end{bmatrix}$. (Alternatively, you may use the SUBTRACT_ELEMENTS function from the VECTOR.MTH file to perform this row operation. See item 4 of Section 5 in Appendix I .) **Author** and **Simplify** `det(#1)` (this is 1 in Figure 5.2; don't confuse it with 1 in Figure 5.1.) The result in expression 3 of Figure 5.2 is the same as the determinant of the original matrix appearing in Figure 5.1 and is independent of t.

$$1: \begin{bmatrix} 4+7t & 2+3t & -3+9t & 5+8t \\ 7 & 3 & 9 & 8 \\ 4 & 5 & 3 & 2 \\ 6 & 6 & 5 & 4 \end{bmatrix}$$

$$2: \text{DET} \begin{bmatrix} 4+7t & 2+3t & -3+9t & 5+8t \\ 7 & 3 & 9 & 8 \\ 4 & 5 & 3 & 2 \\ 6 & 6 & 5 & 4 \end{bmatrix}$$

3: 39

Figure 5.2: **Effect of an elementary row operation on a matrix**

Solved Problem 3: Use Cramer's rule to solve the following system of equations for z.

$$\begin{aligned} 4x + 7y - 8z + 2w &= 9 \\ 3x - y + 2z + 9w &= 7 \\ 5x + 6y + 2z - w &= 3 \\ 8x - 3y + 2z - w &= 5 \end{aligned}$$

Solution: Enter the matrices $\begin{bmatrix} 4 & 7 & -8 & 2 \\ 3 & -1 & 2 & 9 \\ 5 & 6 & 2 & -1 \\ 8 & -3 & 2 & -1 \end{bmatrix}$ and $\begin{bmatrix} 4 & 7 & 9 & 2 \\ 3 & -1 & 7 & 9 \\ 5 & 6 & 3 & -1 \\ 8 & -3 & 5 & -1 \end{bmatrix}$. (You may also use **Declare Matrix** for the first, then use F3 to bring it to the **Author** line and edit the third column. See **Line Editing** in Appendix I.) Now **Author** and **Simplify** `det(#2)/det(#1)`. The result is expression 4 of Figure 5.3.

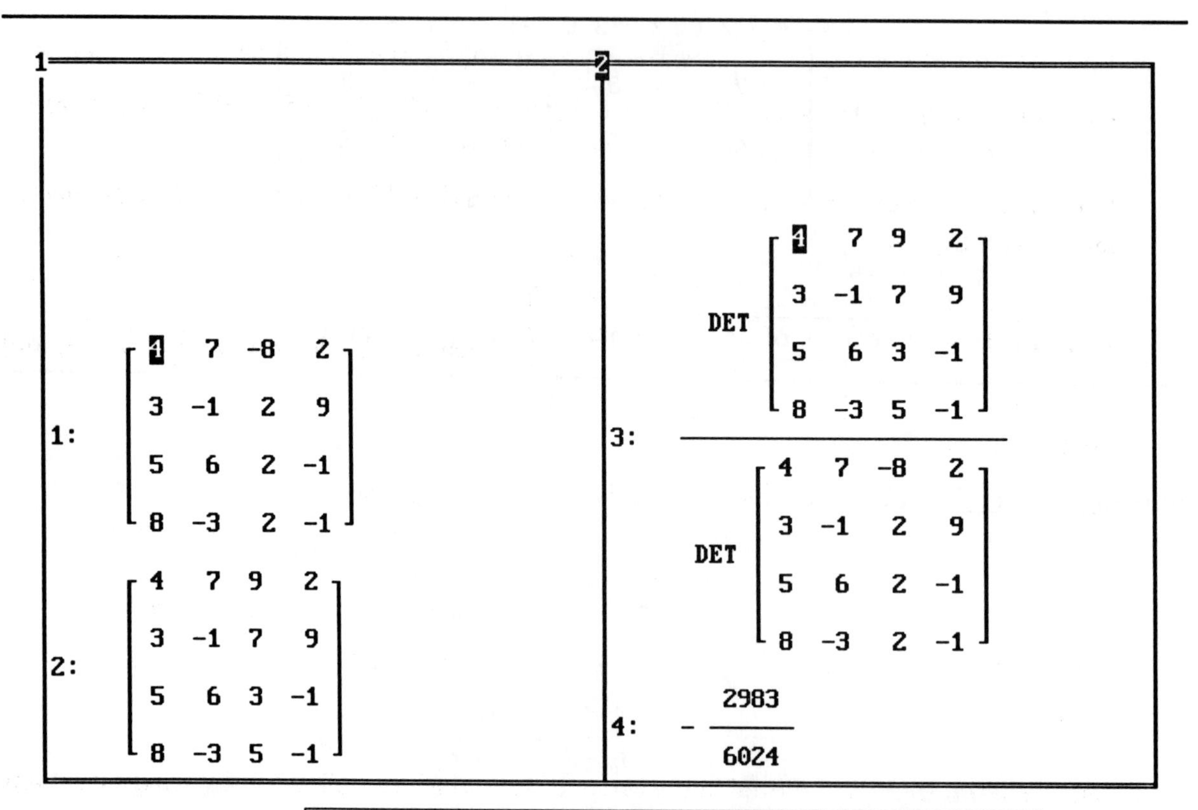

Figure 5.3: Solving a system of equations using Cramer's rule

Exercises

1. Enter the matrix $A = \begin{bmatrix} a & b \\ c & d \end{bmatrix}$. Some books use $|A|$ for the determinant of the matrix A and some use $\text{DET}(A)$. What do you get if you ask *DERIVE* to **Simplify** $|A|$? (Earlier versions of *DERIVE* may not give the same result as version 2.5.)

   Give an example of a *nonzero* 2 by 2 matrix A such that *DERIVE* gives the same answer for $|A|$ and $\text{DET}(A)$.

2. If $A = \begin{bmatrix} 2 & 3 & 5 & 8 \\ 1 & 3 & 2 & 3 \\ 7 & 5 & 4 & 1 \\ 9 & 8 & 1 & 2 \end{bmatrix}$ and $B = \begin{bmatrix} 3 & 6 & 4 & 2 \\ 5 & 4 & 2 & 7 \\ 3 & 9 & 4 & 1 \\ 2 & 4 & 6 & 8 \end{bmatrix}$, calculate the determinants of A, B, and each of the following.

   (a) AB (b) BA (c) $AB - BA$
   (d) $(AB)^2$ (e) A^2B^2 (f) $(AB)^2 - A^2B^2$

   (i) Given your answers to (a) and (b), explain why the answer to (c) is not zero.

   (ii) Given your answers to (d) and (e), explain why the answer to (f) is not zero.

3. Calculate the adjoints of the matrices in exercise 2 above.

4. Use the determinant to determine the values of t for which the following system of equations has a unique solution?

$$\begin{aligned} 2x + 4y + tz &= 3 \\ tx - y + z &= 8 \\ 3x + ty - z &= 12 \end{aligned}$$

5. Find a multiple of the matrix $\begin{bmatrix} 3 & 2 & 9 & 5 \\ 5 & 3 & 7 & 9 \\ 1 & 3 & 4 & 6 \\ 4 & 2 & 3 & 1 \end{bmatrix}$ that has determinant 1.

6. In Chapter 3 we discussed how, given a matrix M, to find a polynomial $p(x)$ such that $p(M) = 0$.

   (a) If $M = \begin{bmatrix} 3 & 2 \\ 6 & 1 \end{bmatrix}$, find $\text{DET}(xI - M)$.

   (b) Your answer above was a polynomial. Evaluate it for $x = M$. (Don't forget that the constant must be multiplied by the identity matrix.)

   (c) Repeat parts (a) and (b) for the matrix in Solved Problem 4 of Chapter 3. Compare the answer you obtain here with that of Solved Problem 4.

7. Repeat parts (a) and (b) of exercise 6 above for the matrix $M = \begin{bmatrix} a & b \\ c & d \end{bmatrix}$.

   Remark: It may seem obvious that if $p(x) = \text{DET}(xI - M)$, then $p(M) = \text{DET}(MI - M) = \text{DET}(0) = 0$. This is a correct result but not a correct argument. See Exploration and Discovery Problem 2(b).

   DERIVE Hint: You must isolate the constant term to multiply it by the identity matrix. It helps to **Expand** $\text{DET}(xI - M)$. *DERIVE* will ask you for variables. Just choose x and press Enter. This will give the polynomial in the standard form $rx^2 + sx + t$.

8. Repeat parts (a) and (b) of exercise 6 above for the matrix $M = \begin{bmatrix} a & b & c \\ d & e & f \\ g & h & i \end{bmatrix}$. See the hints given in exercise 7 above.

9. Use Cramer's rule to solve the following system of equations for b.

$$\begin{aligned} 3a - 4b + 2c - d + 5e &= 3 \\ 8a + 7b - 9c + 2d - e &= 5 \\ 6a - 2b + 3c - 3d + 4e &= 6 \\ 5a + 4b + 6c + 8d + e &= 2 \\ 8a + 2b - 5c + 5d - 9e &= 7 \end{aligned}$$

10. Solve the system in the preceding problem and verify that the solution for b is the same as that given by Cramer's rule.

Exploration and Discovery

1. For each of the following equations, if it is true, prove it. If it is not, find a counter example. Assume that A and B are square matrices in each case.

   (a) $\text{DET}(A+B) = \text{DET}(A) + \text{DET}(B)$

   (b) $\text{DET}(AB) = \text{DET}(A)\text{DET}(B)$

   (c) $\text{DET}(A^n) = \text{DET}(A)^n$

   (d) If A is invertible, then $\text{DET}(A^{-1}) = \dfrac{1}{\text{DET}(A)}$.

   (e) $\text{DET}(cA) = c\,\text{DET}(A)$ for any real number c.

   (f) $\text{DET}(ABA^{-1}) = \text{DET}(B)$

   (g) Let A be an n by n matrix and let B denote the adjoint of A. $\text{DET}(B) = (\det(A))^n$.

2. First you should do Exercises 7 and 8.

   (a) In those two problems we calculated $p(x) = \text{DET}(xI - M)$, evaluated it for $x = M$, and got zero. This important fact is called *The Cayley-Hamilton Theorem.* Verify the Cayley-Hamilton Theorem for 4 by 4 matrices by repeating the procedure in exercises 7 and 8. *Warning: If you name the matrix M using "M:=", be sure not to use the letter "m" as an entry in the matrix.*

   (b) The Cayley-Hamilton Theorem may seem obvious, because of the following "proof."

   $$p(M) = \text{DET}(MI - M) = \text{DET}(\text{zero matrix}) = 0$$

   However, this argument is *not* correct; explain why. Here is a clue: If M is a 2 by 2 matrix, what is $p(M)$ (a number, vector, matrix, ...?) What is $\text{DET}(\text{zero matrix})$ (a number, vector, matrix, ...?)

   (c) Do Laboratory Exercise 3.1. It examines $\text{DET}(A+xB)$ for two choices of B. It is also a polynomial in x. For each choice of B try to discover a matrix C and a constant k such that $\text{DET}(A + xB) = k\text{DET}(xI - C)$. (Hint: Use paper and pencil.) Report your observations and conclusions.

3. Find four 3 by 3 matrices with integer entries (other than triangular matrices) that have determinant 1 or -1. Explain how you arrived at your answers and discuss a strategy for generating such matrices. (Hint: Try letting one or two entries be variables and the rest be numbers.)

4. The matrices in parts (a) and (c) below are known as *Vandermonde matrices.*

   (a) Calculate DET $\begin{bmatrix} 1 & x & x^2 \\ 1 & y & y^2 \\ 1 & z & z^2 \end{bmatrix}$ and **Factor** your answer.

   (b) Based on your answer above, give a necessary and sufficient condition for this determinant to be zero.

   (c) Repeat (a) and (b) for DET $\begin{bmatrix} 1 & x & x^2 & x^3 \\ 1 & y & y^2 & y^3 \\ 1 & z & z^2 & z^3 \\ 1 & w & w^2 & w^3 \end{bmatrix}$.

   (d) Conjecture a factorization for larger Vandermonde matrices and give a necessary and sufficient condition for their det ninants to be zero.

5. $\begin{bmatrix} 1 & \frac{1}{2} \\ \frac{1}{2} & \frac{1}{3} \end{bmatrix}$ and $\begin{bmatrix} 1 & \frac{1}{2} & \frac{1}{3} \\ \frac{1}{2} & \frac{1}{3} & \frac{1}{4} \\ \frac{1}{3} & \frac{1}{4} & \frac{1}{5} \end{bmatrix}$ are known as *Hilbert matrices* of orders 2 and 3 respectively.

To assist in entering a large Hilbert matrix into *DERIVE* , **Author** the following.

```
hilbert(n):=vector(vector(1/(i+j),i,1,n),j,0,n-1)
```

If you now **Author** and **Simplify** `hilbert(n)`, *DERIVE* will return the Hilbert matrix of order n.

   (a) Find the determinants of the Hilbert matrices of orders 2, 3, 5, and 10.

   (b) What do you observe about the magnitude of the determinants of Hilbert matrices?

   (c) Based on your calculations make a conjecture concerning the inevitability of Hilbert matrices. (Try to prove your conjecture.)

   (d) There are many computer programs written in languages such as BASIC, C, Pascal, or Fortran that find inverses and determinants. Unlike *DERIVE,* such programs use "floating point" decimal approximations (such as 0.33333333 for $\frac{1}{3}$), which introduces a small error. What do you think will happen if you use one of these programs to calculate the determinant of a large Hilbert matrix? How about the inverse? If you are able to get access to one, try the 10 by 10 example and report what happens. Try it with *DERIVE* in approximate mode.

6. Calculate the determinant of $A = \begin{bmatrix} 2 & 3 & 5 & 2 \\ 1 & 2 & 1 & 3 \\ 2 & 4 & 1 & 5 \\ 3 & 1 & 2 & 3 \end{bmatrix}$. Change any entry of A by $\frac{1}{100}$ and calculate the determinant again. For example, try $\begin{bmatrix} 2 & 3 & 5 & 2.01 \\ 1 & 2 & 1 & 3 \\ 2 & 4 & 1 & 5 \\ 3 & 1 & 2 & 3 \end{bmatrix}$. How much does the determinant change? Perform this same experiment with other square matrices altering several of the entries by a small amount. Discuss your observations.

Repeat the preceding experiment with the noninvertible matrix $A = \begin{bmatrix} 1 & 2 & 3 \\ 4 & 5 & 6 \\ 7 & 8 & 9 \end{bmatrix}$.

If you select the entries of a square matrix at random, do you expect its determinant to be 0? Do you expect it to be invertible?

Continuation of Problem 6 for students who have studied calculus. Intuitively, a function f is *continuous* if a small change in x produces only a small change in $f(x)$. Discuss the continuity of the determinant as a function of one of its entries.

LABORATORY EXERCISE 5.1

The Determinant of a Sum

Name ______________________________ Due Date ______________

The problem: If A and B are n by n matrices and x is a number, we know that the equations $\mathrm{DET}(A+B) = \mathrm{DET}(A) + \mathrm{DET}(B)$ and $\mathrm{DET}(A+xB) = \mathrm{DET}(A) + x\mathrm{DET}(B)$ are usually false. But if A, B, and x are chosen carefully, we can find cases where these are true. We will investigate two examples. Write your responses to all problems in clear, grammatically correct English sentences.

Let $A = \begin{bmatrix} 3 & 0 & 1 \\ 6 & 1 & 1 \\ 9 & 0 & 7 \end{bmatrix}$, let I be the 3 by 3 identity matrix and let $B = \begin{bmatrix} 1 & 2 & 1 \\ 3 & 4 & 3 \\ -4 & -2 & -4 \end{bmatrix}$.

1. Find $\mathrm{DET}(A)$, $\mathrm{DET}(B)$, $\mathrm{DET}(I)$, $\mathrm{DET}(A + xI)$, and $\mathrm{DET}(A + xB)$.

2. Find all values of x for which $\mathrm{DET}(A + xI) = \mathrm{DET}(A) + \mathrm{DET}(xI)$.

3. Find all values of x for which $\mathrm{DET}(A + xB) = \mathrm{DET}(A) + \mathrm{DET}(xB)$.

4. Compare and contrast your results in the last two problems.

continued on next page.

5. Find all values of x for which $\text{DET}(A + xI) = \text{DET}(A) + x\text{DET}(I)$.

6. Find all values of x for which $\text{DET}(A + xB) = \text{DET}(A) + x\text{DET}(B)$.

7. Compare and contrast your results in the last two problems.

8. Find all values of x such that $A + xI$ is *not* invertible.

9. Find all values of x such that $A + xB$ is *not* invertible.

10. Compare and contrast your results in the last two problems.

CHAPTER 6

APPLICATION
MATRIX ALGEBRA AND MODULAR ARITHMETIC

LINEAR ALGEBRA CONCEPTS

- **Modular Arithmetic**
- **Matrix Operations**
- **Hill Codes**

Introduction

This chapter expands the usual matrix operations using arithmetic modulo a prime. This material is not covered in most linear algebra texts, and it should be considered an optional chapter. However, the laboratory exercise at the end of this chapter contains an interesting application to cryptography called Hill codes.

Notice: If you are using a version of *DERIVE* earlier than 2.5, you will have to use **Transfer Load Utility** MISC to define the MOD function used below. Furthermore, you cannot apply that MOD function to a matrix as we do; you must apply it to each element separately.

A Brief Review of Modular Arithmetic

Let Z_n denote the set $\{0,1,2,\ldots,n-1\}$. If k is any integer, the *residue* of k modulo n is the remainder of k divided by n. If the remainder is r we write $k = r(\text{mod } n)$. For example, $29 = 4(\text{mod } 5)$ because when 29 is divided by 5 we get a remainder of 4. Addition and multiplication in Z_n has the usual definitions except that if the sum or product gets too large, we replace it by its residue modulo n. Thus, in Z_5, $2 \cdot 3 = 6(\text{mod } 5) = 1$ and $3+4 = 7(\text{mod } 5) = 2$.

DERIVE's syntax for the residue of k mod n is MOD(k,n). Thus, if we **Author** and **Simplify** `mod(17,5)`, *DERIVE* will return the value 2 since this is the remainder of 17 divided by 5. *DERIVE* will also perform this operation on matrices: If A is the matrix $\begin{bmatrix} 0 & 1 & 2 \\ 3 & 4 & 5 \\ 6 & 7 & 8 \end{bmatrix}$ and we **Author** and **Simplify** `mod(A,5)`, *DERIVE* will return $\begin{bmatrix} 0 & 1 & 2 \\ 3 & 4 & 0 \\ 1 & 2 & 3 \end{bmatrix}$.

Division modulo an integer n can cause difficulties. For example the obvious solution of the equation $3x = 8(\text{mod } 13)$ is $x = \frac{8}{3}$. But this is not satisfactory, because $\frac{8}{3}$ is not an integer in $\{0, 1, \cdots, 12\}$. The following theorem due to the sixteenth century mathematician Leonhard Euler is needed. Its proof can be found in any elementary number theory text.

Theorem (Euler): If p is prime and k is not a multiple of p, then $k^{p-1} = 1(\text{mod } p)$.

This theorem tells us that $3^{12} = 1(\text{mod } 13)$. *DERIVE* can check this: **Author** and **Simplify** `mod(3^12,13)`. Thus, $\frac{8}{3} = (\frac{8}{3})(1) = (\frac{8}{3})3^{12}(\text{mod } 13)$. If we **Author** and **Simplify** `mod(8/3 3^12,13)` we find that $\frac{8}{3} = 7(\text{mod } 13)$.

In summary, to solve an equation $ax = b(\text{mod } p)$, *calculate the answer* $x = \frac{b}{a}$ *as a rational number and use Euler's theorem to clear the denominator.*

Solved Problems

Solved Problem 1: Solve the following system of equations.

$$\begin{aligned} 3x + 5y - 7z &= 8(\text{mod } 83) \\ 8x - 9y + 13z &= 13(\text{mod } 83) \\ 7x + 4y + 5z &= 15(\text{mod } 83) \end{aligned}$$

Solution: Enter the augmented matrix of the system of equations and ROW_REDUCE. We need to clear the fractions from expression 3 of Figure 6.1. Before applying Euler's theorem you should check that 83 is not a divisor of 701. After this fact has been verified, **Author** and **Simplify** `mod(701^82 #3,83)`. (Since $701^{82} = 1(\text{mod } 83)$, this multiplication does not change the matrix mod 83.) The solution, $x = 50(\text{mod } 83)$, $y = 5(\text{mod } 83)$, and $z = 12(\text{mod } 83)$ can be read from the last column of expression 5.

Solved Problem 2: Find the inverse of $A = \begin{bmatrix} 2 & 17 & 28 \\ 8 & 19 & 2 \\ 21 & 4 & 16 \end{bmatrix}$ modulo 29.

Solution: In this context, the familiar theorem that A has an inverse provided DET$(A) \neq 0$ must be replaced by the theorem that *A has an inverse modulo a prime p if and only if p does not divide* DET(A). Enter the matrix and name it A. Ask *DERIVE* for `det(A)`. Since the determinant, -11146, is not divisible by 29 we conclude that A has an inverse modulo 29.

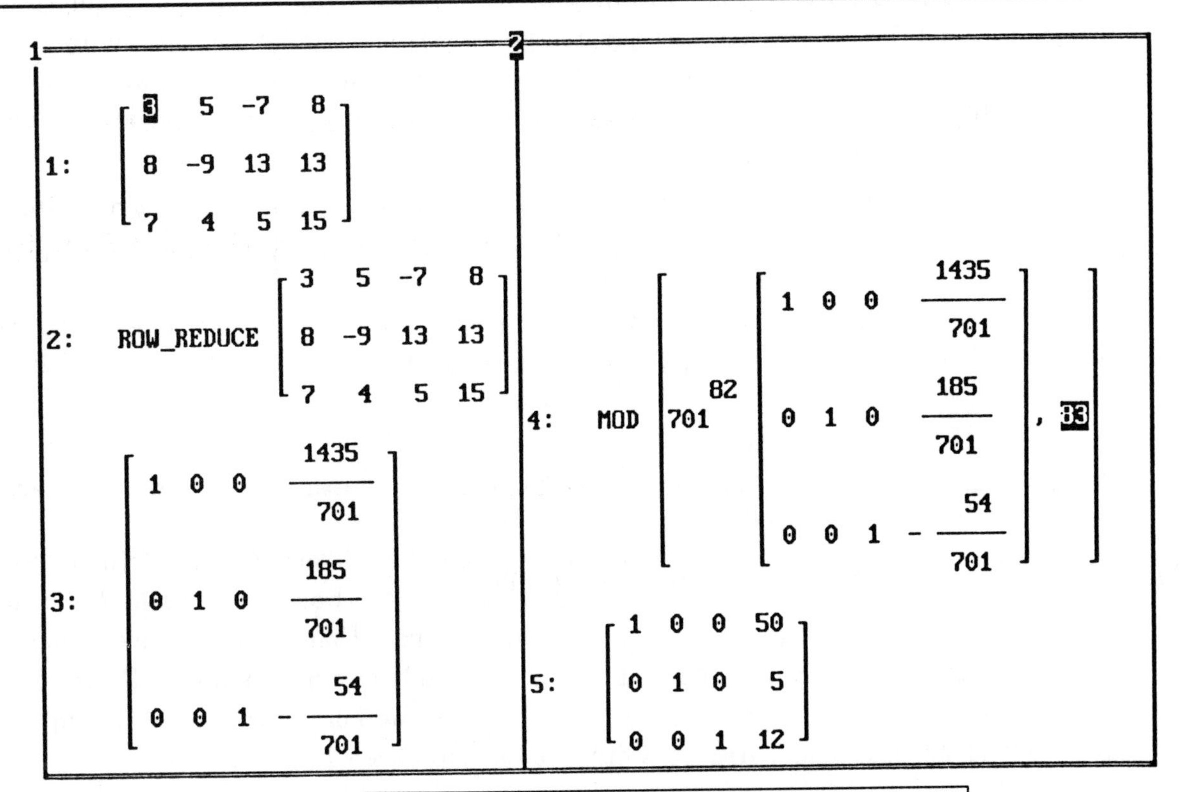

Figure 6.1: Solving a system of equations modulo 83

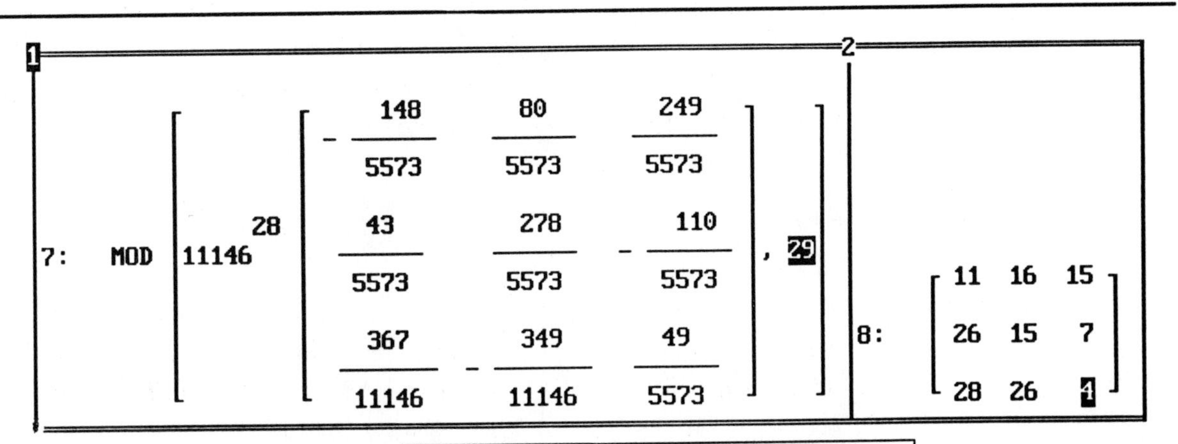

Figure 6.2: The inverse of a matrix modulo 29

The next step is to **Author** and **Simplify** `A^-1`. Notice in expression 7 of Figure 6.2 that $2 \cdot 5573 = 11146$, so we can clear the fractions in expression 8 by multiplying by any power of 11146. Euler's theorem tells us that $11146^{28} = 1 (\text{mod } 29)$. Thus, we **Author** and **Simplify** `mod(11146^28  #8,29)`. The final result is displayed in expression 8. You may wish to check this answer using `mod(A.#8,29)`.

Solved Problem 3: A is a 3 by 3 matrix with integer entries such that $A \begin{bmatrix} 2 \\ 5 \\ 8 \end{bmatrix} = \begin{bmatrix} 3 \\ 9 \\ 2 \end{bmatrix} (\text{mod } 29)$, $A \begin{bmatrix} 4 \\ 6 \\ 1 \end{bmatrix} = \begin{bmatrix} 5 \\ 9 \\ 4 \end{bmatrix} (\text{mod } 29)$, and $A \begin{bmatrix} 8 \\ 7 \\ 6 \end{bmatrix} = \begin{bmatrix} 4 \\ 5 \\ 2 \end{bmatrix} (\text{mod } 29)$. Find $A \begin{bmatrix} 6 \\ 5 \\ 8 \end{bmatrix} (\text{mod } 29)$.

Solution: Let $C = \begin{bmatrix} 2 & 4 & 8 \\ 5 & 6 & 7 \\ 8 & 1 & 6 \end{bmatrix}$ and $D = \begin{bmatrix} 3 & 5 & 4 \\ 9 & 9 & 5 \\ 2 & 4 & 2 \end{bmatrix}$. The three equations above are equivalent to the single matrix equation $AC = D (\text{mod } 29)$. Thus, $A = DC^{-1} (\text{mod } 29)$. Enter these matrices and name them C and D. If you **Author** `det(C)`, *DERIVE* will return -182. This is not divisible by 29, and we conclude that C has an inverse modulo 29. **Author** and **Simplify** `D.C^-1`. From the result in expression 8 of Figure 6.3, we need to multiply by a power of 182 to clear the fractions. **Author** and simplify `mod(182^28 #8,29)`. The matrix A is displayed in expression 10 of Figure 6.3.

Finally, **Author** and **Simplify** `mod(#10 .[6,5,8],29)`. The answer appears in expression 12 of Figure 6.3.

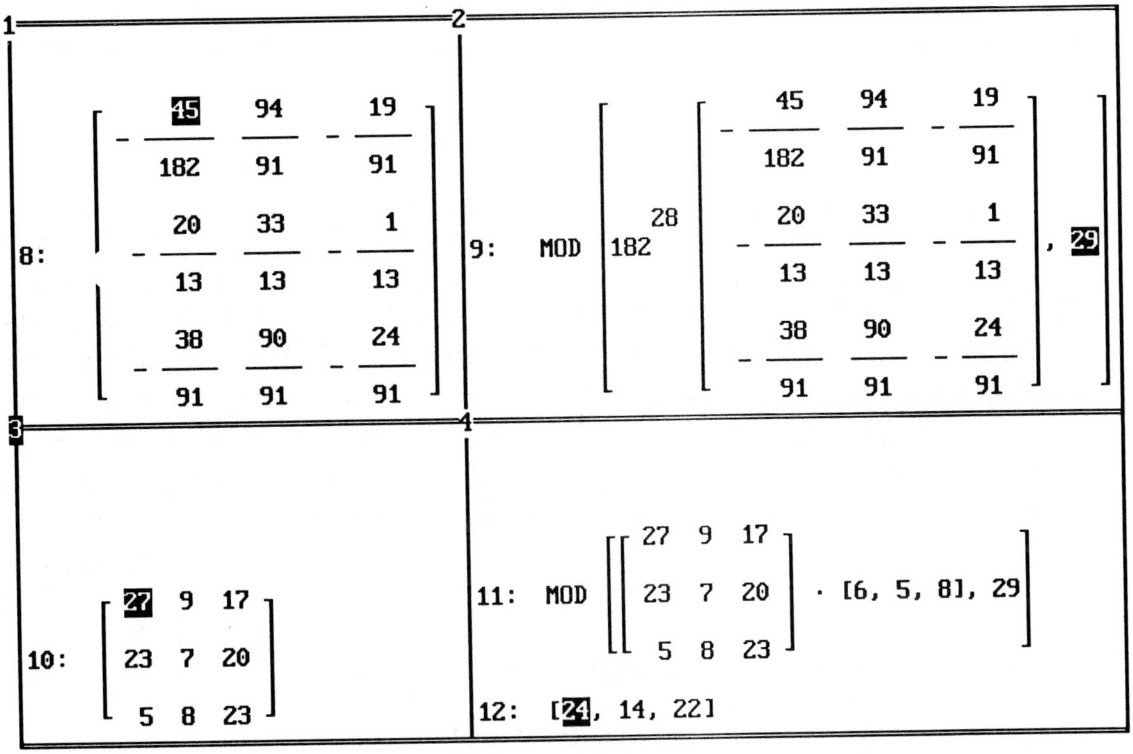

1

8: $$\begin{bmatrix} -\frac{45}{182} & \frac{94}{91} & -\frac{19}{91} \\ -\frac{20}{13} & \frac{33}{13} & -\frac{1}{13} \\ -\frac{38}{91} & \frac{90}{91} & -\frac{24}{91} \end{bmatrix}$$

2

9: $$\text{MOD}\left[182^{28}\begin{bmatrix} -\frac{45}{182} & \frac{94}{91} & -\frac{19}{91} \\ -\frac{20}{13} & \frac{33}{13} & -\frac{1}{13} \\ -\frac{38}{91} & \frac{90}{91} & -\frac{24}{91} \end{bmatrix}, 29\right]$$

3

10: $$\begin{bmatrix} 27 & 9 & 17 \\ 23 & 7 & 20 \\ 5 & 8 & 23 \end{bmatrix}$$

4

11: $$\text{MOD}\left[\begin{bmatrix} 27 & 9 & 17 \\ 23 & 7 & 20 \\ 5 & 8 & 23 \end{bmatrix} \cdot [6, 5, 8], 29\right]$$

12: [24, 14, 22]

Figure 6.3: **A matrix with given vector products**

Exercises

1. Solve the following systems of equations modulo 881.

$$\text{(a)}\quad \begin{aligned} 7x + 4y - 9z &= 18 \\ 3x - 13y + 27z &= 7 \\ 14x - 77y + 38z &= 93 \end{aligned} \qquad \text{(b)}\quad \begin{aligned} 14x + 13y + 6z - 4w &= 17 \\ 12x - 7y + 18z + 4w &= 64 \\ 8x + 4y - 15z + 18w &= 9 \end{aligned}$$

2. For each of the following matrices determine if the inverse modulo 29 exists. If it does, calculate it and check your answer.

$$\text{(a)} \begin{bmatrix} 13 & 4 & 7 \\ 18 & 2 & 9 \\ 3 & 14 & 6 \end{bmatrix} \qquad \text{(b)} \begin{bmatrix} 8 & 7 & 6 \\ 12 & 4 & 11 \\ 2 & 9 & 7 \end{bmatrix}$$

3. Suppose A is a 3 by 3 matrix with integer entries, such that $A \begin{bmatrix} 4 \\ 7 \\ 17 \end{bmatrix} = \begin{bmatrix} 2 \\ 20 \\ 13 \end{bmatrix} \pmod{29}$, $A \begin{bmatrix} 3 \\ 16 \\ 4 \end{bmatrix} = \begin{bmatrix} 2 \\ 18 \\ 9 \end{bmatrix} \pmod{29}$, and $A \begin{bmatrix} 5 \\ 3 \\ 8 \end{bmatrix} = \begin{bmatrix} 2 \\ 17 \\ 23 \end{bmatrix} \pmod{29}$. Find $A \pmod{29}$.

4. Let $A = \begin{bmatrix} 3 & 2 & 5 \\ 4 & 8 & 1 \\ 5 & 4 & 6 \end{bmatrix}$. Find an n such that $A^n = I \pmod{31}$. (Hint: The command `mod(vector(A^n,n,k),31)` will let you view the first k powers of A modulo 31.)

5. If A is an integer matrix and $A^{-1} = \begin{bmatrix} 4 & 6 & 9 \\ 2 & 1 & 4 \\ 19 & 3 & 6 \end{bmatrix} \pmod{37}$, find $A \pmod{37}$.

Exploration and Discovery

1. If A is an integer matrix that has an inverse modulo a prime p, is there always an n such that $A^n = I(\text{mod } p)$? Look at several examples. (Hint: It may help to count how many matrices there are with entries from Z_p.)

2. It can be shown that even if n is not prime, an integer matrix A still has an inverse modulo n provided n and $\text{DET}(A)$ have no common divisors other than 1. That is, $\text{GCD}(n, \text{DET}(A)) = 1$. However, there is a difficulty in calculating the inverse since Euler's theorem is not true if n is not prime.

   (a) Show that $8^{20} \neq 1(\text{mod } 21)$.

   (b) *DERIVE* uses the syntax $\text{GCD}(a, b)$ to denote the greatest common divisor of a and b. Find the greatest common divisors of the pairs $(273, 1491)$, and $(2186, 3294)$.

   (c) The Euler function $\phi(n)$ is defined to be the number of positive integers $k < n$ such that $\text{GCD}(k, n) = 1$. For example $\text{GCD}(1, 6) = 1$, $\text{GCD}(2, 6) = 2$, $\text{GCD}(3, 6) = 3$, $\text{GCD}(4, 6) = 2$, and $\text{GCD}(5, 6) = 1$. Thus $\phi(6) = 2$. Calculate $\phi(26)$ and $\phi(883)$. (Hint. 883 is prime.)

   (d) A second theorem of Euler states that if $\text{GCD}(a, n) = 1$, then $a^{\phi(n)} = 1(\text{mod } n)$. Verify this theorem for $n = 26$ and $a = 17$.

   (e) Find the inverse of $\begin{bmatrix} 2 & 5 & 3 \\ 1 & 4 & 7 \\ 2 & 9 & 2 \end{bmatrix}$ modulo 26.

LABORATORY EXERCISE 6.1

Hill Codes

Name ______________________________ **Due Date** ____________________

One common method of coding is the *Hill code*. (For a more detailed discussion, see *Elementary Linear Algebra, Applications Version* by Howard Anton and Chris Rorres.) The idea is as follows. We add the three symbols , . ? to the alphabet to make the total number prime, and each symbol is assigned a number according to the following table.

A	B	C	D	E	F	G	H	I	J	K	L	M	N	O
0	1	2	3	4	5	6	7	8	9	10	11	12	13	14

P	Q	R	S	T	U	V	W	X	Y	Z	,	.	?
15	16	17	18	19	20	21	22	23	24	25	26	27	28

To illustrate how it works, we will encode **HIDE THIS MESSAGE** using a Hill 3-code with *encryption matrix* $E = \begin{bmatrix} 2 & 7 & 6 \\ 4 & 5 & 13 \\ 2 & 6 & 1 \end{bmatrix}$.

Step 1: Remove the spaces and separate the letters into groups of 3. (Add arbitrary letters to the end if necessary.)

HID ETH ISM ESS AGE

Step 2: Assign each letter its appropriate number from the table above, and arrange these numbers in the columns of a so-called *plaintext* matrix, P as shown:

$$P = \begin{bmatrix} 7 & 4 & 8 & 4 & 0 \\ 8 & 19 & 18 & 18 & 6 \\ 3 & 7 & 12 & 18 & 4 \end{bmatrix}$$

continued on next page

Step 3: Calculate the product EP modulo 29 to obtain the *code matrix:*

$$C = \begin{bmatrix} 1 & 9 & 11 & 10 & 8 \\ 20 & 28 & 17 & 21 & 24 \\ 7 & 13 & 20 & 18 & 11 \end{bmatrix}$$

Step 4: Replace the numbers in the code matrix, C by their corresponding letters.

$$\begin{bmatrix} B & J & L & K & I \\ U & ? & R & V & Y \\ H & N & U & S & L \end{bmatrix}$$

Step 5: Arrange the columns into a single line (without spaces) and send the text:

BUHJ?NLRUKVSIYL

Even if the message is intercepted by an unauthorized person, she will not be able to decode the message unless she has the order of the code (3) and the encryption matrix E. The person to whom we are sending the message has both and can decipher the message by repeating the process above using the inverse modulo 29 of the encryption matrix. Try the following exercises.

1. Calculate the inverse of the encryption matrix E, modulo 29.

Continued on next page.

2. Use your answer to decode the message **BUHJ?NLRUKVSIYL** that we just encoded above. Show your work.

3. Decode **TMB,.PDHJPHV** assuming that it is an order 3 Hill code with the encryption matrix above.

4. You intercept the following message: **GHTIOTZVEKVVXTVB,IZCX**. Naval Intelligence suspects that this is an order 3 Hill code and that the first three words of the message are **I THINK OUR**. Assuming this, find the encryption matrix and decode the message.

CHAPTER 7

VECTOR PRODUCTS, LINES, AND PLANES

LINEAR ALGEBRA CONCEPTS

- **Dot Product**
- **Cross Product**
- **Projection**
- **Unit Vector**
- **Vectors in R^n**
- **Orthogonal vectors**

Introduction

This chapter covers basic operations on vectors in the plane and in 3-space emphasizing the geometric aspects of linear algebra. *DERIVE* 's syntax for the *cross product* of two vectors x and y is CROSS(x, y). The *length* or *norm* of a vector x is denoted by vertical bars, $|x|$.

Notice: In versions of *DERIVE* earlier than 2.5, `|x|` may not give the length of a vector. In this case, you may define your own function by `LEN(x):= sqrt(x.x)`.

Solved Problems

Solved Problem 1: If $\mathbf{u} = (1, 2, 3)$ and $\mathbf{v} = (2, 5, 7)$, find the following.

1. A unit vector in the direction of $\mathbf{u}$
2. $\mathbf{u} \cdot \mathbf{v}$
3. $\mathbf{u} \times \mathbf{v}$
4. The projection of $\mathbf{u}$ onto $\mathbf{v}$

Solution: Each of the items above is easily calculated by hand. The point of this exercise is to introduce *DERIVE* 's syntax for these operations, which will be used later.

The first step is to enter and name the vectors, so we **Author** `u:=[1,2,3]` and `v:=[2,5,7]`.

```
2:   v := [2, 5, 7]

      u
3:   ───
     |u|

      ┌ √14    √14    3 √14 ┐
4:    │ ───,   ───,   ───── │
      └ 14      7      14   ┘

5:   u · v

6:   33

7:   CROSS (u, v)

8:   [-1, -1, 1]

     u · v
9:   ───── v
     v · v

       ┌ 11   55   77 ┐
10:    │ ──,  ──,  ── │
       └ 13   26   26 ┘
```

Figure 7.1: **Basic operations on vectors**

To find a unit vector in the direction of **u**, **Author** and **Simplify** `u/|u|`. The result is expression 4 of Figure 7.1. To get the dot product, **Author** and **Simplify** `u.v`. (The period must be used between **u** and **v**.) The result is expression 6. **Author** and **Simplify** `cross(u,v)` for the cross product in expression 8. To get the projection of **u** onto **v**, **Author** and **Simplify** `(u.v)/(v.v) v`. The result is expression 10 of Figure 7.1.

Solved Problem 2: Find all unit vectors in R^3 that make an angle of 1 radian with both $\mathbf{u} = (1, 2, 3)$ and $\mathbf{v} = (2, 1, 1)$.

Solution: **Author** `u:=[1,2,3]`, `v:=[2,1,1]`, and `w:=[a,b,c]` as seen in Figure 7.2. Assuming **w** is a unit vector, the cosine of the angle between **w** and **u** is $\frac{\mathbf{u}\cdot\mathbf{w}}{|\mathbf{u}|}$. **Author** `[(u.w)/|u|=cos 1, (v.w)/|v|=cos 1]` as seen in expression 4 and **soLve** it using *solve variables a* and *b*. From expression 5 we obtain the following representation for **w**.

$$\mathbf{w} = \left(\frac{(2\sqrt{6}-\sqrt{14})\cos 1 + c}{3}, \frac{(2\sqrt{14}-\sqrt{6})\cos 1 - 5c}{3}, c\right)$$

This is difficult to type in, but we can use the solution in expression 5 to construct **w** in terms of c as follows.

```
1:   u := [1, 2, 3]

2:   v := [2, 1, 1]

3:   w := [a, b, c]
```

$$4: \quad \left[\frac{u \cdot w}{|u|} = \cos(1), \frac{v \cdot w}{|v|} = \cos(1)\right]$$

$$5: \quad \left[a = \left[\frac{2\sqrt{6}}{3} - \frac{\sqrt{14}}{3}\right]\cos(1) + \frac{c}{3}, b = \left[\frac{2\sqrt{14}}{3} - \frac{\sqrt{6}}{3}\right]\cos(1) - \frac{5c}{3}\right]$$

$$6: \quad \left[\left[\frac{2\sqrt{6}}{3} - \frac{\sqrt{14}}{3}\right]\cos(1) + \frac{c}{3}, \left[\frac{2\sqrt{14}}{3} - \frac{\sqrt{6}}{3}\right]\cos(1) - \frac{5c}{3}, c\right]$$

$$7: \quad \left|\left[\left[\frac{2\sqrt{6}}{3} - \frac{\sqrt{14}}{3}\right]\cos(1) + \frac{c}{3}, \left[\frac{2\sqrt{14}}{3} - \frac{\sqrt{6}}{3}\right]\cos(1) - \frac{5c}{3}, c\right]\right| = 1$$

Figure 7.2: **A vector making equal angles with two others. Part I**

Highlight expression 5 with the arrow keys; press A for **Author** and then F3. This brings the expression to the **Author** line for editing. Using Ctrl S and Ctrl D, move the cursor on the author line and delete "a =" and "b =" from the expression; then insert ",c" at the end before the square bracket "]". (See Section 8 in Appendix I for more details on editing.) Press Enter and you should see expression 6 in Figure 7.2.

Now we must find the values of c that make **w** a unit vector. **Author** and **soLve** `|#8|=1`, to get the solutions in expressions 8 and 9 of Figure 7.3. We see that there are two solutions. To finish the problem, we **approX**imate these numbers and substitute them back into expression 6 of Figure 7.2. Highlight expression 6 and use **Manage Substitute**. When you are asked for a substitute value for c, highlight the value of c in expression 10 of in Figure 7.3 by pressing the right arrow key twice; then press F3. (You must delete the "c" from the end before pressing Enter.) Expression 12 is far too complicated to be

$$8: \quad c = \left[\frac{11\sqrt{14}}{35} - \frac{\sqrt{6}}{5}\right] \text{COS}(1) - \frac{3\sqrt{7}\sqrt{(4\sqrt{3}(\sqrt{7} - 2\sqrt{3})\text{COS}(1)^2 + 5)}}{35}$$

$$9: \quad c = \frac{3\sqrt{7}\sqrt{(4\sqrt{3}(\sqrt{7} - 2\sqrt{3})\text{COS}(1)^2 + 5)}}{35} + \left[\frac{11\sqrt{14}}{35} - \frac{\sqrt{6}}{5}\right] \text{COS}(1)$$

$$10: \quad c = -0.0440796$$

$$11: \quad c = 0.785430$$

$$12: \quad \left[\left[\frac{2\sqrt{6}}{3} - \frac{\sqrt{14}}{3}\right] \text{COS}(1) + \frac{-0.0440796}{3}, \left[\frac{2\sqrt{14}}{3} - \frac{\sqrt{6}}{3}\right] \text{COS}(1) - \frac{5\ (-0}{}\right.$$

$$13: \quad [0.193741, 0.980061, -0.0440796]$$

$$14: \quad \left[\left[\frac{2\sqrt{6}}{3} - \frac{\sqrt{14}}{3}\right] \text{COS}(1) + \frac{0.78543}{3}, \left[\frac{2\sqrt{14}}{3} - \frac{\sqrt{6}}{3}\right] \text{COS}(1) - \frac{5\ 0.7854}{3}\right.$$

$$15: \quad [0.470244, -0.402454, 0.785429]$$

Figure 7.3: **A vector making equal angles with two others. Part II**

useful. (In Figure 7.3 expressions 12 and 14 go off the screen.) The **approX** command returns an approximation for **w** in expression 13.

Repeat this procedure using the second solution for c in expression 13 to obtain the second vector in expression 15.

Notice that, geometrically, the set of all vectors that make an angle of 1 radian with a fixed vector forms a cone. Thus, we are looking for unit vectors that lie on the intersection of two cones with a common vertex.

Solved Problem 3: Find a vector **w** whose projections onto $\mathbf{u} = (1, 2, 3)$, $\mathbf{v} = (2, 1, 1)$, and $\mathbf{x} = (3, 1, 2)$ are $(2, 4, 6)$, $(4, 2, 2)$, and $(6, 2, 4)$ respectively.

Solution: First **Author** each of the following as a separate expression:

```
w:=[a,b,c]          u:=[1,2,3]          v:=[2,1,1]          x:=[3,1,2]
```

1: w := [a, b, c]

2: u := [1, 2, 3]

3: v := [2, 1, 1]

4: x := [3, 1, 2]

5: $$\left[\frac{w \cdot u}{u \cdot u} u = 2 u, \frac{w \cdot v}{v \cdot v} v = 2 v, \frac{w \cdot x}{x \cdot x} x = 2 x\right]$$

6: $$\begin{bmatrix} \frac{a + 2 b + 3 c}{14} = 2 & \frac{a + 2 b + 3 c}{7} = 4 & \frac{3 (a + 2 b + 3 c)}{14} = 6 \\ \frac{2 a + b + c}{3} = 4 & \frac{2 a + b + c}{6} = 2 & \frac{2 a + b + c}{6} = 2 \\ \frac{3 (3 a + b + 2 c)}{14} = 6 & \frac{3 a + b + 2 c}{14} = 2 & \frac{3 a + b + 2 c}{7} = 4 \end{bmatrix}$$

Figure 7.4: **A vector with three given projections: Part I**

Notice that the projection of **w** onto **u** is 2**u**, the projection of **w** onto **v** is 2**v**, and the projection of **w** onto **x** is 2**x**. Thus we **Author** and **Simplify**

```
[(w.u)/(u.u) u=2u,(w.v)/(v.v) v=2v,(w.x)/(x.x) x=2x]
```

DERIVE presents the system of nine equations in three unknowns in expression 6 of Figure 7.4. **Author** and **Simplify** append F4 to list the equations in vector form as is seen in expression 8 of Figure 7.5. Finally, **soLve** the system in expression 8 to obtain the solution from expression 9 of Figure 7.5, $\mathbf{w} = (3, -7, 13)$.

```
7:  APPEND [ (a + 2 b + 3 c)/14 = 2     (a + 2 b + 3 c)/7 = 4     3 (a + 2 b + 3 c)/14
             (2 a + b + c)/3 = 4        (2 a + b + c)/6 = 2       (2 a + b + c)/6 = 2
             3 (3 a + b + 2 c)/14 = 6   (3 a + b + 2 c)/14 = 2    (3 a + b + 2 c)/7 =

8:  [(a + 2 b + 3 c)/14 = 2, (a + 2 b + 3 c)/7 = 4, 3 (a + 2 b + 3 c)/14 = 6, (2 a +)/3

9:  [a = 3, b = -7, c = 13]
```

Figure 7.5: **A vector with three given projections: Part II**

<u>**Solved Problem 4**</u>**:** Find all planes that pass through the points $(4,1,5)$ and $(2,5,3)$ and whose distance from the point $(1,2,4)$ is 2.

<u>Solution</u>: It is left to the reader to show that the required plane does not pass through the origin. (Hint: Find the equation of the plane through $(4,1,5)$, $(2,5,3)$, and $(0,0,0)$, and show that its distance from $(1,2,4)$ is not 2.) Thus its equation can be written in the form $ax + by + cz + 1 = 0$. (Explain why.)

It is a fact that the distance from the point (x_0, y_0, z_0) to the plane $ax + by + cz + d = 0$ is $\frac{|ax_0 + by_0 + cz_0 + d|}{|(a,b,c)|}$. Therefore, the information given in the problem tells us the following.

$$\begin{aligned} 4a + b + 5c + 1 &= 0 \\ 2a + 5b + 3c + 1 &= 0 \\ \frac{|a + 2b + 4c + 1|}{|(a,b,c)|} &= 2 \end{aligned}$$

The first step is to **Author** and **soLve** the system of equations

```
[4a+b+5c+1=0,2a+5b+3c+1=0]
```

$$1:\quad [4a + b + 5c + 1 = 0,\ 2a + 5b + 3c + 1 = 0]$$

$$2:\quad \left[a = -\frac{11c + 2}{9},\ b = -\frac{c + 1}{9}\right]$$

$$3:\quad \frac{|a + 2b + 4c + 1|}{|[a, b, c]|} - 2$$

$$4:\quad \frac{\left|-\frac{11c + 2}{9} + 2\left[-\frac{c + 1}{9}\right] + 4c + 1\right|}{\left|\left[-\frac{11c + 2}{9}, -\frac{c + 1}{9}, c\right]\right|} - 2$$

$$5:\quad \frac{|23c + 5|}{\sqrt{(203c^2 + 46c + 5)}} - 2$$

Figure 7.6: **Planes through two points a distance 2 from a third point: Part I**

using a and b as *solve variables.* The solution is expression 2 of Figure 7.6. Next **Author** `|a+2b+4c+1|/|[a,b,c]|-2` and use **Manage Substitute** to replace a and b by their representations in terms of c in expression 2. **Simplify** to obtain expression 5 of Figure 7.6.

Now **soLve** to get the zeros as expressions 6 and 7 of Figure 7.7. **approX** these to obtain the solutions in expressions 8 and 9 of Figure 7.7. If we plug these values of c back into expression 2 using **Manage Substitute**, we obtain the solutions as expressions 11 and 13. Thus, the required planes are

$$\begin{aligned} -0.131135x - 0.102830y - 0.0745258z + 1 &= 0 \\ -0.511974x - 0.137452y + 0.237070z + 1 &= 0 \end{aligned}$$

6: $c = \frac{23}{283} - \frac{18\sqrt{6}}{283}$

7: $c = \frac{18\sqrt{6}}{283} + \frac{23}{283}$

8: c = -0.0745258

9: c = 0.237070

10: $\left[a = -\frac{11\ (-0.0745258) + 2}{9},\ b = -\frac{-0.0745258 + 1}{9}\right]$

11: [a = -0.131135, b = -0.102830]

12: $\left[a = -\frac{11\ 0.23707 + 2}{9},\ b = -\frac{0.23707 + 1}{9}\right]$

13: [a = -0.511974, b = -0.137452]

Figure 7.7: **Planes through two points a distance 2 from a third point: Part II**

<u>**Solved Problem 5**</u>: Find all vectors of length 2 in R^4 that are orthogonal to all three of $(1,5,2,8)$, $(3,5,7,9)$, and $(6,2,1,3)$.

<u>Solution</u>: First **Author** each of the following as a separate expression:

`u:=[a,b,c,d]`   `x:=[1,5,2,8]`   `y:=[3,5,7,9]`   `z:=[6,2,1,3]`

Just as in lower dimensions, vectors are orthogonal provided that their dot product is zero. Thus we **Author** and **soLve** `[u.x=0,u.y=0,u.z=0]`. Using a, b and c as *solve variables*, we obtain the solution as in expression 6 of Figure 7.8. Since the vectors we seek have length 2, we **Author** and **soLve** `|d[1/23,-35/23,-5/23,1]|=2`. The solutions appear in expressions 8 and 9. Thus we get the vectors $\pm\frac{23}{\sqrt{445}}\left(\frac{1}{23}, -\frac{35}{23}, -\frac{5}{23}, 1\right)$.

```
1:   u := [a, b, c, d]

2:   x := [1, 5, 2, 8]

3:   y := [3, 5, 7, 9]

4:   z := [6, 2, 1, 3]

5:   [u · x = 0, u · y = 0, u · z = 0]

          d            35 d          5 d
6:   [a = ——, b = - ———————, c = - ————]
          23            23            23

        1       35       5
7:   |d [——, - ——, - ——, 1]| = 2
        23      23       23

          23 √445
8:   d = - ———————
            445

         23 √445
9:   d = ———————
           445
```

Figure 7.8: **Vectors in R^4 orthogonal to 3 given vectors**

Solved Problem 6: **For students who have studied calculus.**

1. Establish the product rule for differentiation of dot products. That is,

$$\text{if } f(t) = [x(t), y(t)] \text{ and } g(t) = [z(t), w(t)] \text{ then}$$
$$\frac{d}{dt}\{f(t) \cdot g(t)\} = f'(t) \cdot g(t) + f(t) \cdot g'(t)$$

2. Show how it follows quickly from (1) that $\frac{d}{dt}|f(t)|^2 = 2f(t) \cdot f'(t)$.
3. Show that the tip of the vector $\mathbf{v}$ is nearest the curve $g(t)$ at a point where the tangent vector to $g(t)$ is orthogonal to $\mathbf{v}-g(t)$.
4. Find the point on the curve $g(t) = (t, \frac{1}{t})$ that is nearest $(1, 0)$.

Solution: Parts 1 – 3 do not require a computer; however, see Exploration and Discovery Problem 2(h).

1.
$$\begin{aligned}\frac{d}{dt}\{f(t)\cdot g(t)\} &= \frac{d}{dt}\{x(t)z(t)+y(t)w(t)\}\\ &= x'(t)z(t)+x(t)z'(t)+y'(t)w(t)+y(t)w'(t)\\ &= [x'(t),y'(t)]\cdot[z(t),w(t)]+[x(t),y(t)]\cdot[z'(t),w'(t)]\\ &= f'(t)\cdot g(t)+f(t)\cdot g'(t)\end{aligned}$$

2. Observe that $|f(t)|^2 = f(t)\cdot f(t)$. Thus, setting $g(t) = f(t)$ in part (1) gives $\frac{d}{dt}|f(t)|^2 = \frac{d}{dt}(f(t)\cdot f(t)) = f'(t)\cdot f(t)+f(t)\cdot f'(t) = 2f(t)\cdot f'(t)$.

3. The distance from the tip of $\mathbf{v}$ to a point $g(t)$ on the curve is the length of the vector $\mathbf{v}-g(t)$. To minimize this distance, it is sufficient to minimize its square, $|\mathbf{v}-g(t)|^2$. This minimum occurs where the derivative is 0, so we have to solve $\frac{d}{dt}|\mathbf{v}-g(t)|^2 = 0$. According to part (2) above we can calculate the derivative as follows:

$$\frac{d}{dt}|\mathbf{v}-g(t)|^2 = 2(\mathbf{v}-g(t))\cdot\frac{d}{dt}(\mathbf{v}-g(t)) = -2(\mathbf{v}-g(t))\cdot g'(t)$$

Thus the critical points occur where $(\mathbf{v}-g(t))\cdot g'(t) = 0$, that is, where $(\mathbf{v}-g(t))$ is orthogonal to the tangent vector $g'(t)$.

4. We'll apply part 3 with $\mathbf{v} = [0,1]$.

Author `g(t):=[t,1/t]` and `([1,0]-g(t)).dif(g(t),t)`.

If we **Simplify** we obtain expression 3 of Figure 7.9 which must be solved. (Recall that solving an expression in *DERIVE* is the same as finding its zeros.) *DERIVE* cannot solve this exactly so we must solve it approximately. (Refer to Section 11 of Appendix I for details.) Plot the graph of expression 3 as in window 2 of Figure 7.9. From the graph it is clear that the point we seek is between $t = 1$ and $t = 2$. Set *DERIVE* to its approximate mode using **Options Precision Approximate**, and ask *DERIVE* to **soLve** on the interval [1, 2]. The solution is displayed as expression 4. **Author** and **approX** `g(1.38027)` to obtain the answer displayed in expression 6.

As we can see from the graph there is also a negative solution in the interval [−1, 0]. We leave it to the reader to find it and to interpret its meaning.

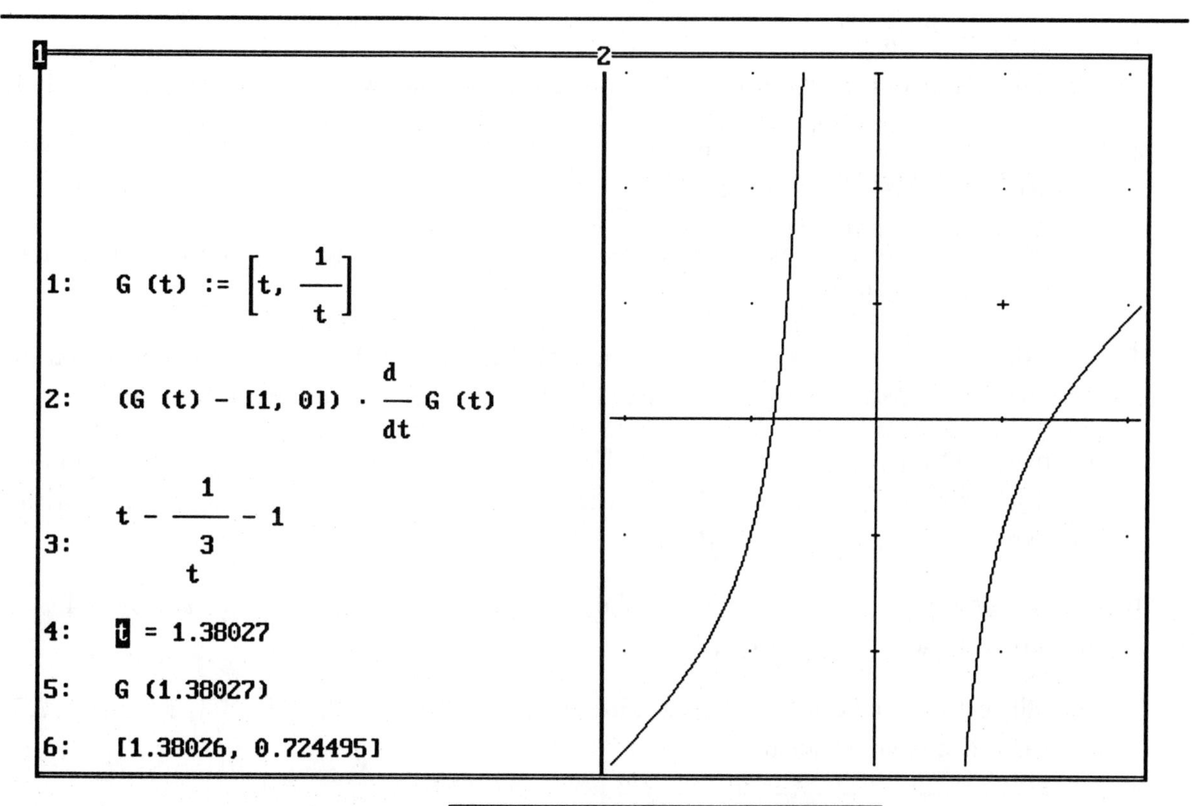

Figure 7.9: A point on $(t, \frac{1}{t})$ nearest $(1, 0)$

Exercises

1. Find all vectors of length 2 that make an angle of 1 radian with both $(1,3,1)$ and $(2,1,3)$.

2. Find a vector whose projections onto $(1,2,3)$, $(2,1,4)$, and $(3,1,1)$ are, respectively, $(3,6,9)$, $(-2,-1,-4)$, and $(6,2,2)$.

3. Find the equations of all planes that pass through the points $(3,2,1)$ and $(4,1,1)$ and are a distance 3 from $(6,2,7)$.

4. Let L denote the line through the origin and the point $(1,2,3)$. Find parametric equations of all lines that pass through the point $(7,2,1)$ and meet L at an angle of 1 radian.

5. **Author** `x:=[a,b,c]` and `y:=[r,s,t]`. Use *DERIVE* to verify that $\mathbf{x} \times \mathbf{y}$ is orthogonal to $\mathbf{x}$ and $\mathbf{y}$. Explain what you did. (*DERIVE* can be used to verify other facts and identities. See Exploration and Discovery Problem 2.)

6. Find all points on the intersection of the planes $3x + 7y - 9z = 8$ and $2x - 3y + 4z = 5$ that are a distance 6 from $(1,2,3)$.

7. Find all vectors of length 3 in R^5 that are orthogonal to $(3,6,4,2,1)$, $(2,6,4,2,1)$, $(5,1,2,7,7)$, and $(9,1,2,5,7)$ simultaneously.

8. **For students who have studied calculus.** Find the point on the curve $(t, \ln t)$ that is nearest the point $(1,1)$.

Exploration and Discovery

1. Parts (a) and (b) below do not require a computer.

   (a) If (a,b,c), (d,e,f), and (g,h,i) are given points, show that $\text{DET}\begin{bmatrix} 1 & x & y & z \\ 1 & a & b & c \\ 1 & d & e & f \\ 1 & g & h & i \end{bmatrix} = 0$ is the equation of a plane.

   (b) Show that the plane in part (a) passes through the points (a,b,c), (d,e,f), and (g,h,i). (You should be able to do this by substitution without expanding the determinant.)

   (c) Use this method to find the equation of the plane through the points $(1,4,9)$, $(8,7,2)$, and $(3,3,7)$.

   (d) Use this method to find the equation of the plane through the points $(1,3,7)$, $(2,5,9)$ and $(3,8,16)$. Explain what happens and why.

   (e) Show that $\text{DET}\begin{bmatrix} x & y & z \\ a & b & c \\ d & e & f \end{bmatrix}$ is the equation of a plane that passes through the points (a,b,c) and (d,e,f). Prove that it must also pass through the origin.

2. *DERIVE* can be used to verify identities. **Author** `x:=[a,b,c]` and `y:=[r,s,t]`.

   (a) **Author** and **Expand** `|x-y|^2 -|x|^2 -|y|^2`. **Author** and **Expand** `x.y`. What relationship do you discover between $|\mathbf{x}-\mathbf{y}|^2 - |\mathbf{x}|^2 - |\mathbf{y}|^2$ and $\mathbf{x}\cdot\mathbf{y}$?

   (b) **Author** and **Expand** `|x+y|^2 +|x-y|^2`. How is it related to $|\mathbf{x}|$ and $|\mathbf{y}|$?

   (c) How is $\left(\frac{\mathbf{x}\cdot\mathbf{y}}{|\mathbf{x}||\mathbf{y}|}\right)^2$ related to $\left(\frac{|\mathbf{x}\times\mathbf{y}|}{|\mathbf{x}||\mathbf{y}|}\right)^2$?

   (d) Using $\mathbf{x}\cdot\mathbf{y} = |\mathbf{x}||\mathbf{y}|\cos\theta$ and the preceding identity you discovered, express $|\mathbf{x}\times\mathbf{y}|$ in terms of $|\mathbf{x}|$, $|\mathbf{y}|$, and the angle θ. Be sure to explain how your answer follows from the previous identity (c).

   (e) Try to discover other identities by this method.

(f) **Author** `x:=[a,b,c,d]` and `y:=[r,s,t,u]`. Try to verify identities (a), (b), and (c) that you discovered above for $\mathbf{R}^4$. If anything strange happens, explain why.

(g) **Author** `x:=[a,b]` and `y:=[r,s]`. Try to verify identities (a), (b) and (c) you discovered above for $\mathbf{R}^2$. If anything strange happens, explain why.

(h) In Solved Problem 6 we discussed several results involving calculus. It may seem a good idea to let *DERIVE* do them. Let's look at $\frac{d}{dt}|f(t)|^2 = f(t) \cdot f'(t)$. **Author** each of the following four expressions separately.

`x(t):=`   `y(t):=`   `f(t):=[x(t), y(t)]`   `|f(t)|^2.`

With the last expression highlighted, choose **Calculus Differentiate** and press Enter until $\frac{d}{dt}|f(t)|^2$ appears. **Simplify** it. Does it look like the result we got in Solved Problem 6 part 3? Explain the differences and why they occur. (In *DERIVE*, SIGN(x) is 1 if $x > 0$, -1 if $x < 0$, and undefined if $x = 0$.)

LABORATORY EXERCISE 7.1

Angles Between Vectors

Name ______________________ **Due Date** ______________

<u>The Problem</u>: Let $\mathbf{x} = [3, 2, 1]$ and $\mathbf{y} = [1, 2, 7]$. Follow the ideas in Solved Problem 1 to answer the following questions.

1. How many degrees are in an angle of one radian?

2. Find all *unit* vectors that make an angle of one radian with both $\mathbf{x}$ and $\mathbf{y}$.

3. Check your answers in (2) above.

4. Find all *unit* vectors that make an angle of one-half a radian with both $\mathbf{x}$ and $\mathbf{y}$. *(This problem has a pitfall. You should explain what difficulty you encounter here and why.)*

continued on next page

5. Describe geometrically the set of all vectors of the form $[1, 2, a]$.

6. Find all vectors of the form $[1, 2, a]$. (no restriction on the length) that make an angle of one radian with $\mathbf{x}$.

CHAPTER 8

VECTOR SPACES AND SUBSPACES

LINEAR ALGEBRA CONCEPTS

- **Vector space**
- **Subspace**
- **Spaces of functions and matrices**
- **Linear combination**
- **Spanning set**
- **Null space**
- **Rank of a matrix**

Introduction

Many problems concerning general vector spaces can be reduced to questions about systems of linear equations, which *DERIVE* can help solve. We will use R^n to denote Euclidean n-space, P_n the space of polynomials of degree no greater than n, and $M_{p,q}$ the space of p by q matrices.

Solved Problems

Solved Problem 1: Let S denote the subspace of R^4 spanned by the vectors $\mathbf{u} = (1,2,3,4)$, $\mathbf{v} = (4,2,1,5)$ and $\mathbf{w} = (3,5,1,7)$. Determine if the vectors $\mathbf{x} = (8,9,5,16)$ and $\mathbf{y} = (7,2,1,3)$ are in S.

Solution: **Author** `u:=[1,2,3,4]`, `v:=[4,2,1,5]`, `w:=[3,5,1,7]`, `x:=[8,9,5,16]` and `y:=[7,2,1,3]` as seen in Figure 8.1. We need to determine if **x** and **y** are linear combinations of the vectors **u**, **v**, and **w**.

Author and **Simplify** `x=au+bv+cw`. If we ask *DERIVE* to **soLve** this system of equations, we deduce from expression 8 of Figure 8.1 that $\mathbf{x} = \mathbf{u} + \mathbf{v} + \mathbf{w}$. Thus, **x** is in S. Repeat this procedure for the equation `y=au+bv+cw`. We see from the status line in the lower left corner of the screen (Figure 8.1) that this equation has no solution. Thus, **y** is not in S.

```
1:   u := [1, 2, 3, 4]

2:   v := [4, 2, 1, 5]

3:   w := [3, 5, 1, 7]

4:   x := [8, 9, 5, 16]

5:   y := [7, 2, 1, 3]

6:   x = a u + b v + c w

7:   [8 = a + 4 b + 3 c, 9 = 2 a + 2 b + 5 c, 5 = 3 a + b + c, 16 = 4 a + 5 b +

8:   [a = 1, b = 1, c = 1]

9:   y = a u + b v + c w

10:  [7 = a + 4 b + 3 c, 2 = 2 a + 2 b + 5 c, 1 = 3 a + b + c, 3 = 4 a + 5 b + 7

COMMAND: Author Build Calculus Declare Expand Factor Help Jump soLve Manage
         Options Plot Quit Remove Simplify Transfer moVe Window approX
No solutions found
Simp(9)                E:LIST.MTH            Free:100%            Derive Algebra
```

Figure 8.1: The span of three vectors in R^4

Alternative solution: We can answer both questions by row-reducing the matrix in expression 2 of Figure 8.2. This is an augmented matrix with the first three columns representing the coefficient matrix. The first four columns of the row-reduced form (expression 3 of Figure 8.2) give us the solution we got in expression 8 of Figure 8.1. The first three columns along with the last one in the row-reduced form show that there is no solution in the second case. Be sure you understand why this alternative solution works.

$$2: \quad \text{ROW\_REDUCE} \begin{bmatrix} 1 & 4 & 3 & 8 & 7 \\ 2 & 2 & 5 & 9 & 2 \\ 3 & 1 & 1 & 5 & 1 \\ 4 & 5 & 7 & 16 & 3 \end{bmatrix}$$

$$3: \quad \begin{bmatrix} 1 & 0 & 0 & 1 & 0 \\ 0 & 1 & 0 & 1 & 0 \\ 0 & 0 & 1 & 1 & 0 \\ 0 & 0 & 0 & 0 & 1 \end{bmatrix}$$

Figure 8.2: **The span of three vectors in R^4: Alternative solution**

Solved Problem 2: Find a vector in R^5 that is not in the subspace spanned by the vectors $(1,2,3,2,1)$, $(3,2,7,1,6)$, $(2,5,6,1,4)$, and $(4,1,7,2,2)$.

Solution: We seek a vector $\mathbf{v} = (p,q,r,s,t)$ so that the following equation has no solution.

$$a(1,2,3,2,1) + b(3,2,7,1,6) + c(2,5,6,1,4) + d(4,1,7,2,2) = \mathbf{v}$$

Author and **Simplify**

```
a[1,2,3,2,1]+b[3,2,7,1,6]+c[2,5,6,1,4]+d[4,1,7,2,2]=[p,q,r,s,t]
```

(As we've suggested before, rather than **Author** the whole thing, you may want to **Author** each vector on a separate line and assemble them into an equation by referring to them by expression number. See Section 3 in Appendix I .) When we ask *DERIVE* to **soLve** it will prompt us to choose four solve variables. In Figure 8.3 we have accepted the default solve variables, a, b, c, d and p. In expression 3 we see $p = \dfrac{q - 3r + s + t}{4}$. Thus, any vector $\mathbf{v}$ that does not satisfy this equation will do. We use $\mathbf{v} = (0,0,0,0,1)$ as an example.

Alternative solution using the file REDUCE.MTH in Appendix II. Load the file REDUCE.MTH using **Transfer Merge** REDUCE. The augmented matrix of the system of equations is

```
1:   a [1, 2, 3, 2, 1] + b [3, 2, 7, 1, 6] + c [2, 5, 6, 1, 4] + d [4, 1, 7, 2,

2:   [a + 3 b + 2 c + 4 d = p, 2 a + 2 b + 5 c + d = q, 3 a + 7 b + 6 c + 7 d =

     - 43 s - 19 t           11 q - 37 r + 27 s + 35 t           q - 3 r + s + t ┐
3:   ─────────────, d = - ─────────────────────────────, p = - ───────────────── ]
     24                                 124                             4        ┘
```

Figure 8.3: **A vector not in a given subspace**

$$A = \left[\begin{array}{cccc|c} 1 & 3 & 2 & 4 & p \\ 2 & 2 & 5 & 1 & q \\ 3 & 7 & 6 & 7 & r \\ 2 & 1 & 1 & 2 & s \\ 1 & 6 & 4 & 2 & t \end{array}\right]$$

Use **Declare Matrix** to enter this 5 by 5 matrix as expression 1. Now **Author** and **Simplify** `reduce(#1,4)` to get the result in Figure 8.4. From the last row of the matrix in Figure 8.4 we see that the system of equations has a solution if and only if $p + \frac{q-3r+s+t}{4} = 0$, and we arrive at the previous solution.

<u>**Solved Problem 3**</u>: Determine if the vectors $(2,1,3,5)$, $(4,3,7,2)$, $(3,1,2,5)$, $(2,1,5,6)$, and $(4,1,3,1)$ span R^4.

<u>Solution</u>: These five vectors span R^4 if and only if the following equation has a solution for each vector $\mathbf{v}$ in R^4.

$$a(2,1,3,5) + b(4,3,7,2) + c(3,1,2,5) + d(2,1,5,6) + e(4,1,3,1) = \mathbf{v}$$

Author and **Simplify** the following:

```
a[2,1,3,5]+b[4,3,7,2]+c[3,1,2,5]+d[2,1,5,6]+e[4,1,3,1]=[p,q,r,s]
```

(All the expressions in Figure 8.5 go off the screen. Use Ctrl ← and Ctrl → to scroll across.) As we've suggested before, you may want to **Author** each vector on a separate line and refer to them by number in the equation rather than **Author** the whole thing.

$$
12: \begin{bmatrix}
1 & 0 & 0 & 0 & \frac{q}{31} - \frac{9 r}{31} + \frac{25 s}{31} + \frac{6 t}{31} \\
0 & 1 & 0 & 0 & -\frac{19 q}{124} - \frac{15 r}{124} + \frac{21 s}{124} + \frac{41 t}{124} \\
0 & 0 & 1 & 0 & \frac{33 q}{124} + \frac{13 r}{124} - \frac{43 s}{124} - \frac{19 t}{124} \\
0 & 0 & 0 & 1 & -\frac{11 q}{124} + \frac{37 r}{124} - \frac{27 s}{124} - \frac{35 t}{124} \\
0 & 0 & 0 & 0 & p + \frac{q}{4} - \frac{3 r}{4} + \frac{s}{4} + \frac{t}{4}
\end{bmatrix}
$$

Figure 8.4: Alternative solution to Solved Problem 2

```
1:   a [2, 1, 3, 5] + b [4, 3, 7, 2] + c [3, 1, 2, 5] + d [2, 1, 5, 6] + e [4, 1

2:   [2 a + 4 b + 3 c + 2 d + 4 e = p, a + 3 b + c + d + e = q, 3 a + 7 b + 2 c
```

$$
3: \quad \left[a = \frac{57 e - 20 p + 43 q - 9 r + 7 s}{11}, \; b = -\frac{10 e - p - 5 q - r + 2 s}{22}, \; c \right.
$$

Figure 8.5: A spanning set for R^4

Ask *DERIVE* to **soLve** this, accepting the default *solve variables.* We see from scrolling expression 3 that the solution does not place any restrictions on p, q, r, or s. We conclude that there is a solution no matter what their values are, and so the five given vectors span R^4.

Alternative solution using the *rank* of a matrix. The vector equation above is equivalent to the following matrix equation.

$$\begin{bmatrix} 2 & 4 & 3 & 2 & 4 \\ 1 & 3 & 1 & 1 & 1 \\ 3 & 7 & 2 & 5 & 3 \\ 5 & 2 & 5 & 6 & 1 \end{bmatrix} \begin{bmatrix} a \\ b \\ c \\ d \\ e \end{bmatrix} = \begin{bmatrix} p \\ q \\ r \\ s \end{bmatrix}$$

$$2: \quad \text{ROW\_REDUCE} \begin{bmatrix} 2 & 4 & 3 & 2 & 4 \\ 1 & 3 & 1 & 1 & 1 \\ 3 & 7 & 2 & 5 & 3 \\ 5 & 2 & 5 & 6 & 1 \end{bmatrix}$$

$$3: \quad \begin{bmatrix} 1 & 0 & 0 & 0 & -\frac{57}{11} \\ 0 & 1 & 0 & 0 & \frac{5}{11} \\ 0 & 0 & 1 & 0 & \frac{32}{11} \\ 0 & 0 & 0 & 1 & \frac{21}{11} \end{bmatrix}$$

Figure 8.6: **Alternative method of showing that vectors span R^4**

Enter the coefficient matrix and ROW_REDUCE it. The result is expression 3 of Figure 8.6. Notice that there is a "1" in each row of the reduced form of the coefficient matrix. This ensures that the system has a solution. We conclude once more that the five given vectors span R^4.

The crucial observation in the work above is that each row of the reduced form of the coefficient matrix contains a leading 1. This leads to the following:

Definition: The *rank* of a matrix A is the number of leading "1s" (or the number of nonzero rows) in the reduced echelon form of A.

The rank is easily calculated using *DERIVE*. Simply ROW_REDUCE and count the number of nonzero rows. The above observation yields the following at once:

Theorem: Let A be a matrix with n rows. The columns of A span R^n if and only if the rank of A is n.

To determine if a set of vectors spans R^n, arrange them as columns of a matrix. The vectors span R^n if and only if the rank of this matrix is n.

Solved Problem 4: Let S denote the subspace of $M_{2,2}$ consisting of matrices of the form $\begin{bmatrix} a+b+c & a-c \\ 3b+3c & 2a-b+4c \end{bmatrix}$, and let T denote the subspace of $M_{2,2}$ consisting of matrices of the form $\begin{bmatrix} p-q+3r & p+2q \\ r-q & p-q-r \end{bmatrix}$. Describe all matrices that are in both S and T.

Solution: Use **Declare Matrix** to enter the 2 by 2 matrices above as expressions 1 and 2 (not shown in Figure 8.7). We wish to find choices of a, b, c and p, q, r that make these two matrices the same. **Author #1=#2** as in expression 3 in Figure 8.7 and **Simplify**. To change the matrix of equations in expression 4 into a vector, **Author** and **Simplify** **append(#4)**. Next **soLve** the system of equations in expression 6 of Figure 8.7 using solve variables a, b, c and p. From the last entry in expression 7, we see that the original equation has a solution if and only if $p = \dfrac{2(20q-29r)}{3}$. Highlight expression 2 and use **Manage Substitute** to substitute this value for p. Those vectors that are in both subspaces are of the form displayed in expression 8 of Figure 8.7. It can be simplified further, but we haven't shown it. The matrices we have described form the intersection of two subspaces, which should again be a subspace. The description shows this.

Solved Problem 5: Let $p(x) = x^3 + x^2 + 1$, $q(x) = 2x^3 - x^2 - 3x + 1$, and $r(x) = x^3 - x + 4$, and let S denote the subspace of P_3 spanned by $p(x)$, $q(x)$, and $r(x)$. For what values of a, b, c, and d does $ax^3 + bx^2 + cx + d$ belong to S?

Solution: First **Author** as separate expressions

`p(x):=x^3+x^2+1,`   `q(x):=2x^3-x^2-3x+1,`   and   `r(x):=x^3-x+4`

3: $$\begin{bmatrix} a + b + c & a - c \\ 3 b + 3 c & 2 a - b + 4 c \end{bmatrix} = \begin{bmatrix} p - q + 3 r & p + 2 q \\ r - q & p - q - r \end{bmatrix}$$

4: $$\begin{bmatrix} a + b + c = p - q + 3 r & a - c = p + 2 q \\ 3 b + 3 c = r - q & 2 a - b + 4 c = p - q - r \end{bmatrix}$$

5: $$\text{APPEND} \begin{bmatrix} a + b + c = p - q + 3 r & a - c = p + 2 q \\ 3 b + 3 c = r - q & 2 a - b + 4 c = p - q - r \end{bmatrix}$$

6: $$[a + b + c = p - q + 3 r, a - c = p + 2 q, 3 b + 3 c = r - q, 2 a - b + 4 c$$

7: $$\frac{2 (19 q - 25 r)}{3}, b = \frac{7 (q - r)}{3}, c = \frac{8 (r - q)}{3}, p = \frac{2 (20 q - 29 r)}{3}]$$

8: $$\begin{bmatrix} \frac{2 (20 q - 29 r)}{3} - q + 3 r & \frac{2 (20 q - 29 r)}{3} + 2 q \\ r - q & \frac{2 (20 q - 29 r)}{3} - q - r \end{bmatrix}$$

Figure 8.7: **The intersection of subspaces of $M_{2,2}$**

We wish to know what restrictions must be placed on a, b, c, and d to ensure that we can solve $ax^3 + bx^2 + cx + d = ep(x) + fq(x) + gr(x)$ for e, f, and g. **Author** and **Simplify** `ax^3+bx^2+cx+d=ep(x)+fq(x)+gr(x)`. The result appears in expression 5 of Figure 8.8. Equating coefficients yields the following system of equations.

$$\begin{aligned} a &= e + 2f + g \\ b &= e - f \\ c &= -(3f + g) \\ d &= e + f + 4g \end{aligned}$$

This system of equations appears as a list in expression 6 of Figure 8.8. *DERIVE* cannot **soLve** this system of equations using only the *solve variables* e, f, and g. (This indicates that some restrictions must be placed on a, b, c, and d to guarantee a solution.) The solution using *solve variables* e, f, g, and a appears in expression 7 of Figure 8.8 (part of

1: $P(x) := x^3 + x^2 + 1$

2: $Q(x) := 2x^3 - x^2 - 3x + 1$

3: $R(x) := x^3 - x + 4$

4: $ax^3 + bx^2 + cx + d = e\,P(x) + f\,Q(x) + g\,R(x)$

5: $ax^3 + bx^2 + cx + d = x^3 (e + 2f + g) + x^2 (e - f) - x(3f + g) + e + f$

6: $[a = e + 2f + g,\ b = e - f,\ c = -(3f + g),\ d = e + f + 4g]$

7: $= \dfrac{11b - 4c - d}{10},\ f = \dfrac{b - 4c - d}{10},\ g = -\dfrac{3b - 2c - 3d}{10},\ a = b - c]$

Figure 8.8: **A subspace of P_3**

it runs off the left of the screen). We conclude from expression 7 that a solution exists provided $a = b - c$. Thus, polynomials in S are of the form $(b - c)x^3 + bx^2 + cx + d$.

Solved Problem 6: Find a spanning set for the null space of the matrix $A = \begin{bmatrix} 2 & 7 & 8 & 4 & 1 \\ 3 & 2 & 5 & 1 & 9 \\ 4 & 1 & 3 & 8 & 1 \end{bmatrix}$.

Solution: The null space of A is the collection of all solutions of the equation $A\mathbf{x} = \mathbf{0}$. Enter the matrix and ROW_REDUCE it. From expression 3 of Figure 8.9 we see that the solution is

$$x_1 = -\tfrac{11}{3}x_4 + \tfrac{97}{39}x_5 \qquad x_2 = -\tfrac{10}{3}x_4 + \tfrac{209}{39}x_5 \qquad x_3 = \tfrac{10}{3}x_4 - \tfrac{212}{39}x_5$$

1: $\begin{bmatrix} 2 & 7 & 8 & 4 & 1 \\ 3 & 2 & 5 & 1 & 9 \\ 4 & 1 & 3 & 8 & 1 \end{bmatrix}$

2: ROW_REDUCE $\begin{bmatrix} 2 & 7 & 8 & 4 & 1 \\ 3 & 2 & 5 & 1 & 9 \\ 4 & 1 & 3 & 8 & 1 \end{bmatrix}$

3: $\begin{bmatrix} 1 & 0 & 0 & \frac{11}{3} & -\frac{97}{39} \\ 0 & 1 & 0 & \frac{10}{3} & -\frac{209}{39} \\ 0 & 0 & 1 & -\frac{10}{3} & \frac{212}{39} \end{bmatrix}$

Figure 8.9: **A spanning set for the null space of a matrix**

$$\text{Thus, } \mathbf{x} = \begin{bmatrix} -\frac{11}{3}x_4 + \frac{97}{39}x_5 \\ -\frac{10}{3}x_4 + \frac{209}{39}x_5 \\ \frac{10}{3}x_4 - \frac{212}{39}x_5 \\ x_4 \\ x_5 \end{bmatrix} = x_4 \begin{bmatrix} -\frac{11}{3} \\ -\frac{10}{3} \\ \frac{10}{3} \\ 1 \\ 0 \end{bmatrix} + x_5 \begin{bmatrix} \frac{97}{39} \\ \frac{209}{39} \\ -\frac{212}{39} \\ 0 \\ 1 \end{bmatrix}$$

This says that each vector in the null space of A is a linear combination of the vectors $\begin{bmatrix} -\frac{11}{3} \\ -\frac{10}{3} \\ \frac{10}{3} \\ 1 \\ 0 \end{bmatrix}$ and $\begin{bmatrix} \frac{97}{39} \\ \frac{209}{39} \\ -\frac{212}{39} \\ 0 \\ 1 \end{bmatrix}$. Notice that the first vector is obtained by setting $x_4 = 1$ and $x_5 = 0$ and that the second is obtained by setting $x_4 = 0$ and $x_5 = 1$. Thus, these two vectors belong to the null space of A and form a spanning set. (Note: It is wise to check your work by multiplying the two vectors by A to see if you do indeed get the zero vector. We leave this to the reader.)

This exercise indicates a procedure for finding a spanning set for the null space of a matrix.

Procedure for Constructing a Spanning Set for the Null Space of a Matrix

1. ROW_REDUCE the matrix to obtain the solution of $A\mathbf{x} = \mathbf{0}$.
2. Obtain a solution corresponding to each independent variable by setting that variable to 1 and the other independent variables to 0.
3. The collection of vectors obtained in (2) form a spanning set for the null space.

Exercises

1. Let S denote the subspace of R^5 spanned by $(2,3,1,6,4)$, $(3,5,1,2,3)$, $(4,2,1,5,7)$, and $(3,8,2,2,1)$.

   (a) Determine if $(6,2,1,11,13)$ is in S.

   (b) For what values of a, b, c, d, and e does (a,b,c,d,e) belong to S?

   (c) Find the intersection of S with the subspace of R^5 spanned by $(2,9,1,8,8)$, $(4,1,6,7,7)$, and $(3,5,7,9,1)$.

   (d) **This exercise uses the file REDUCE.MTH in Appendix II..** Find a vector in R^5 that is not in S.

2. Let U denote the subspace of P_4 spanned by the polynomials $p(x) = x^4 - 3x^3 + x^2 - x + 1$, $r(x) = 2x^4 - 3x^2 + x + 3$, $s(x) = x^4 + x^3 + x^2 + 4x - 1$, and $t(x) = 3x^4 + x^3 - 2x^2 + 3$.

   (a) Determine if $x^4 - 3x^3 + 4x^2 + 2x - 7$ is in U.

   (b) For what values of a, b, c, d, and e does $ax^4 + bx^3 + cx^2 + dx + e$ belong to U?

3. Let T be the subspace of $M_{3,3}$ consisting of matrices of the form $\begin{bmatrix} a & b & c \\ a-b & a+c & b-2c \\ a+b+c & a+3c & 2a+b \end{bmatrix}$.
   Find the intersection of T with the subspace of $M_{3,3}$ consisting of matrices of the form $\begin{bmatrix} a & b & a+b \\ c & a+c & d \\ e & c+d+e & f \end{bmatrix}$.

4. **This exercise may be solved using rank.** Determine if the following sets of vectors span R^5.

   (a) $(1,3,2,5,6)$, $(3,3,2,1,5)$, $(5,6,8,1,2)$, $(3,6,5,1,8)$, $(5,5,2,6,7)$

   (b) $(1,3,2,1,1)$, $(2,1,1,3,1)$, $(1,1,2,2,3)$, $(3,4,3,4,2)$, $(3,2,3,5,4)$

5. **This exercise may be solved using rank.** Determine if the matrices $\begin{bmatrix} 3 & 7 \\ 2 & 5 \end{bmatrix}$, $\begin{bmatrix} 5 & 9 \\ 2 & 3 \end{bmatrix}$, $\begin{bmatrix} 6 & 6 \\ 1 & 2 \end{bmatrix}$, $\begin{bmatrix} 4 & 5 \\ 9 & 1 \end{bmatrix}$ span $M_{2,2}$.

6. Find spanning sets for the null spaces of the following matrices.

(a) $\begin{bmatrix} 3 & 5 & 9 & 8 \\ 2 & 3 & 1 & 7 \\ 4 & 1 & 2 & 9 \\ 6 & 3 & 4 & 1 \end{bmatrix}$ (b) $\begin{bmatrix} 2 & 3 & 4 & 1 & 2 \\ 5 & 2 & 4 & 8 & 1 \\ 3 & 1 & 2 & 5 & 4 \\ 8 & 3 & 6 & 13 & 5 \end{bmatrix}$

7. If $f(x)$ is a function and r is a number, a new function can be defined by $f(x - r)$. It is called a *translate* of f. Let $f(x) = x^2$. If r, s, and t are distinct real numbers, show that $\{f(x - r), f(x - s), f(x - t)\}$ is a spanning set for P_2.

LABORATORY EXERCISE 8.1

Describing Sets of Vectors

Name ______________________ **Due Date** ______________

Let $A = \begin{bmatrix} 3 & 5 & 9 & 8 \\ 2 & 3 & 1 & 7 \\ 4 & 1 & 2 & 9 \\ 6 & 3 & 4 & 1 \end{bmatrix}$. Let S denote the subspace of R^4 spanned by $(3, 2, 1, 1)$, $(1, 1, 3, 1)$, $(2, 1, -2, 0)$, and $(4, 3, 4, 2)$.

1. Find a spanning set for the null space of A.

2. Find all vectors that are in both S and the null space of A.

CHAPTER 9

INDEPENDENCE, BASIS, AND DIMENSION

LINEAR ALGEBRA CONCEPTS

- **Linearly independent set**
- **Basis**
- **Dimension**
- **Coordinate vector**

Introduction

A vector space has many spanning sets. A spanning set that is "minimal" in the sense that it does not contain unnecessary vectors is a *basis*. The number of vectors in a basis is called the *dimension* of the vector space.

An element $\mathbf{x}$ in a vector space with a basis S can be written as a linear combination of the elements of S in a unique way, provided that we acknowledge that the elements of S are in a prescribed order. When the coefficients in this linear combination are written as a vector, it is called the coordinate vector of $\mathbf{x}$ relative to the *ordered* basis S. When we discuss the coordinate vector of $\mathbf{x}$ relative to a basis S, we will always assume S is ordered even though ordinary set notation is used.

For example, the coordinate vector of the polynomial $x^2 - x$ relative to the ordered basis $\{x^2, x, 1\}$ is $(1, -1, 0)$, and the coordinate vector of $x^2 - x$ relative to the ordered basis $\{x, 1, x^2\}$ is $(-1, 0, 1)$.

Solved Problems

Solved Problem 1: Determine if the following vectors are linearly independent: (1,4,6,8,7,2), (5,2,9,1,3,5), (3,6,8,1,4,7), (5,5,2,8,6,3).

Solution: **Author** the following:

```
u:=[1,4,6,8,7,2], v:=[5,2,9,1,3,5], w:=[3,6,8,1,4,7], x:=[5,5,2,8,6,3]
```

```
1:   u := [1, 4, 6, 8, 7, 2]

2:   v := [5, 2, 9, 1, 3, 5]

3:   w := [3, 6, 8, 1, 4, 7]

4:   x := [5, 5, 2, 8, 6, 3]

5:   a u + b v + c w + d x = [0, 0, 0, 0, 0, 0]

6:   [a + 5 b + 3 c + 5 d = 0, 4 a + 2 b + 6 c + 5 d = 0, 6 a + 9 b + 8 c + 2 d

7:   [a = 0, b = 0, c = 0, d = 0]
```

Figure 9.1: **Linearly independent vectors in R^6**

(See Figure 9.1.) We want to determine if there is a unique solution to $a\mathbf{u}+b\mathbf{v}+c\mathbf{w}+d\mathbf{x} = 0$. Recall that if *DERIVE* is asked to **soLve** an expression that is not an equation, it sets the expression equal to zero before solving. If the expression is a vector, *DERIVE* sets it equal to the zero vector. Thus, you may **Author** and **soLve** `au+bv+cw+dx`. (Alternatively you may enter the vectors without naming them and enter this equation referring to expression numbers. See Solved Problem 2.) *DERIVE* Warning: It is common practice to write $a\mathbf{u} + b\mathbf{v} + c\mathbf{w} + d\mathbf{x} = 0$ and assume the reader will know from the context that 0 must be the zero vector in R^6 and not the number 0. It's the only way it makes sense. But *DERIVE* interprets 0 as a scalar rather than the zero vector, so we must be careful. If you wish to enter the equation in this way you must use `au+bv+cw+dx=[0,0,0,0,0,0]` as in expression 5 of Figure 9.1.

Alternative Solution: Another method for determining if vectors in R^n are linearly independent is to make a matrix using the given vectors as the columns. Of course, this is just the coefficient matrix of the system we solved previously. *The vectors are linearly independent if and only if the rank of this matrix is the same as the number of vectors.*

Use **Declare Matrix** to enter the 6 by 4 matrix $\begin{bmatrix} 1 & 5 & 3 & 5 \\ 4 & 2 & 6 & 5 \\ 6 & 9 & 8 & 2 \\ 8 & 1 & 1 & 8 \\ 7 & 3 & 4 & 6 \\ 2 & 5 & 7 & 3 \end{bmatrix}$. Now ask *DERIVE* to `row_reduce(#1)` and **Simplify**.

From expression 3 of Figure 9.2 we see that the rank of the matrix is the same as the number of vectors (4), and we conclude once more that the vectors are linearly independent.

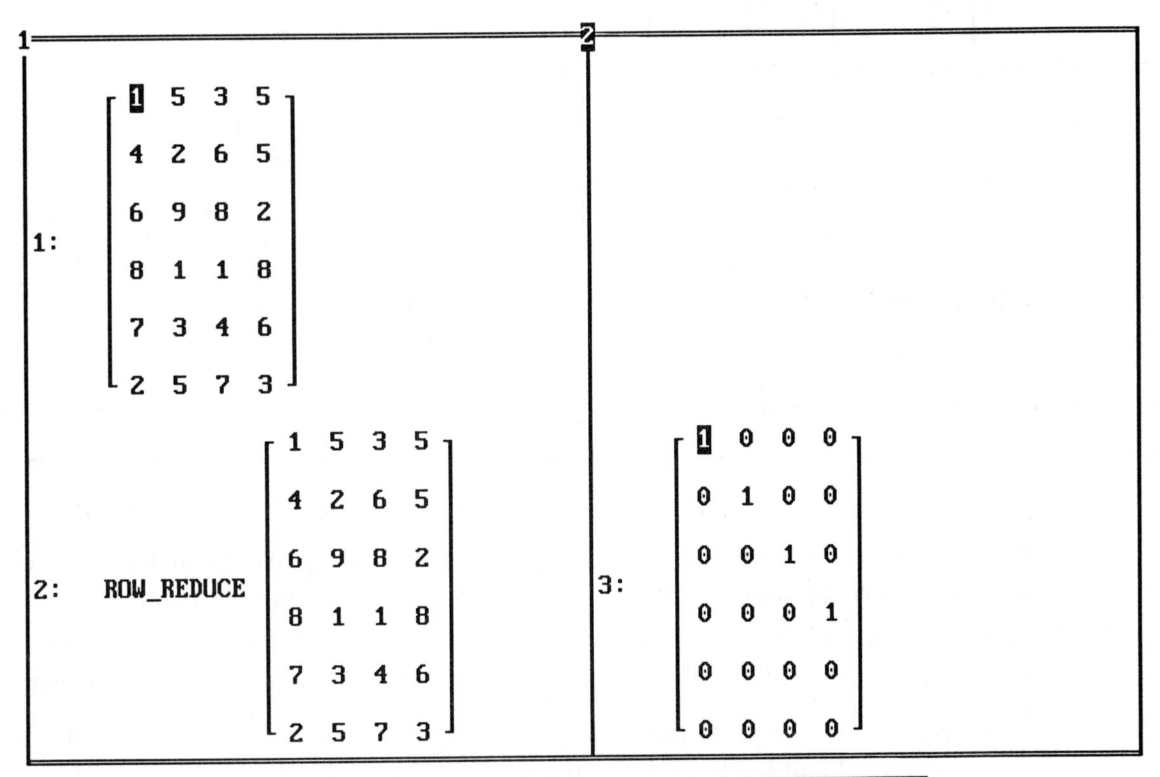

Figure 9.2: **Test for independence using the rank**

Solved Problem 2: Determine if the following elements of $M_{3,2}$ are linearly independent. If they are linearly dependent, write one as a linear combination of the others.

$$\mathbf{U} = \begin{bmatrix} 2 & 3 \\ 1 & 4 \\ 3 & 2 \\ 1 & 5 \end{bmatrix} \quad \mathbf{V} = \begin{bmatrix} 1 & 5 \\ 4 & 2 \\ 3 & 1 \\ 2 & 2 \end{bmatrix} \quad \mathbf{W} = \begin{bmatrix} 3 & 8 \\ 5 & 6 \\ 6 & 3 \\ 3 & 7 \end{bmatrix} \quad \mathbf{X} = \begin{bmatrix} 1 & -2 \\ -3 & 2 \\ 0 & 1 \\ -1 & 3 \end{bmatrix}$$

Solution: Use **Declare Matrix** to enter each of the matrices separately. Assuming they are expressions 1 through 4, we want to set `a#1 +b#2 +c#3 +d#4` equal to the zero matrix and solve for a,b,c, and d. But recall that it must first be converted to a *vector* of equations before *DERIVE* can solve it. This is done as usual with the APPEND function. **Author** `append(a#1 +b#2 +c#3 +d#4)` and then **soLve** as seen in Figure 9.3. (We do not need

```
5:   APPEND [a [2 3; 1 4; 3 2; 1 5] + b [1 5; 4 2; 3 1; 2 2] + c [3 8; 5 6; 6 3; 3 7] + d [1 -2; -3 2; 0 1; -1 3]]

6:   [2 a + b + 3 c + d, 3 a + 5 b + 8 c - 2 d, a + 4 b + 5 c - 3 d, 4 a + 2 b +

7:   [a = @1, b = @2, c = - (@1 + @2)/2, d = (@2 - @1)/2]
```

$$\text{APPEND}\left[a\begin{bmatrix}2&3\\1&4\\3&2\\1&5\end{bmatrix}+b\begin{bmatrix}1&5\\4&2\\3&1\\2&2\end{bmatrix}+c\begin{bmatrix}3&8\\5&6\\6&3\\3&7\end{bmatrix}+d\begin{bmatrix}1&-2\\-3&2\\0&1\\-1&3\end{bmatrix}\right]$$

Figure 9.3: **Testing elements in $M_{3,2}$ for independence**

to set it equal to zero. *DERIVE* assumes it. If you do set the equation equal to zero, you must use the 4 by 2 zero matrix. See the *DERIVE* warning in Solved Problem 1.) We see from expression 7 of Figure 9.3 that there are infinitely many solutions dependent on arbitrary constants @1 and @2. We conclude that the elements are linearly dependent.

The second part of this problem has infinitely many correct solutions. If we choose the parameter @1 to be 2 and @2 to be 0, we obtain $2\mathbf{U} - \mathbf{W} - \mathbf{X} = 0$ so that $\mathbf{X} = 2\mathbf{U} - \mathbf{V}$. Any assignment of values for the parameters @1 and @2 will lead to a correct solution, except @1 = @2 = 0. (What's wrong with this choice?)

Solved Problem 3: In $C(-\infty,\infty)$ determine if the functions e^x, $\sin x$, $\cos x$, $\ln(x^2+1)$ are linearly independent.

Solution: We wish to determine if there are any nontrivial solutions of

$$ae^x + b\sin x + c\cos x + d\ln(x^2+1) = 0$$

It is important to remember here that the zero in this equation is the constant function $f(x) = 0$ rather than the real number zero. Thus we want to see if there are values of a, b, c, and d, not all zero, that make $ae^x + b\sin x + c\cos x + d\ln(x^2+1)$ zero for *all* x. If it is zero for *all* x, it must be zero for any four particular values of x: for example, $x = 0$, $x = 1$, $x = 2$, and $x = 3$. Why did we choose these four? Just because they are convenient. More about this at the end of the problem. We now seek solutions of the following system of equations.

$$
\begin{aligned}
a+c &= 0 \\
ae+b\sin 1+c\cos 1+d\ln 2 &= 0 \\
ae^2+b\sin 2+c\cos 2+d\ln 5 &= 0 \\
ae^3+b\sin 3+c\cos 3+d\ln 10 &= 0
\end{aligned}
$$

It is a little tedious to enter these equations into *DERIVE*, so here's a way that saves some typing. (See Figure 9.4.) **Author** `f(x):=a` Alt e `^x +b sin x +c cos x +d ln x`. We can now **Author** the whole system of equations as `[f(0), f(1), f(2), f(3)]`. **Simplify** and **soLve** to obtain the solution displayed in expression 4 of Figure 9.4. We conclude that the four functions are linearly independent.

DERIVE Warning: To enter the function e^x, we must use Alt e `^x` as we did above. *DERIVE* interprets e as a variable just as it would a or b. Notice in expression 1 of Figure 9.4 that *DERIVE* has put a circumflex on top of the letter e to distinguish the number e from the variable e.

We chose $x = 0, 1, 2, 3$ because they are convenient, but we could have picked any four distinct values and they would probably have led us to the same conclusion. However, it is possible that the resulting system may have a nontrivial solution. What would you conclude then? Be careful. (See Exploration and Discovery Problem 1(b).)

```
                 x                                      2
1:   F (x) := a ê  + b SIN (x) + c COS (x) + d LN (x  + 1)

2:   [F (0), F (1), F (2), F (3)]

                                                         2
3:   [a + c, ê a + d LN (2) + c COS (1) + b SIN (1), ê  a + d LN (5) + c COS (2)

4:   [a = 0, b = 0, c = 0, d = 0]
```

Figure 9.4: **Independence of functions in $C(-\infty,\infty)$**

Solved Problem 4: Show that the set

$$S = \{(2,6,3,4,2),\ (3,1,5,8,3),\ (5,1,2,6,7),\ (8,4,3,2,6),\ (5,5,6,3,4)\}$$

is a basis for R^5 and find the coordinate vector of $(3,4,1,7,8)$ relative to the ordered basis S.

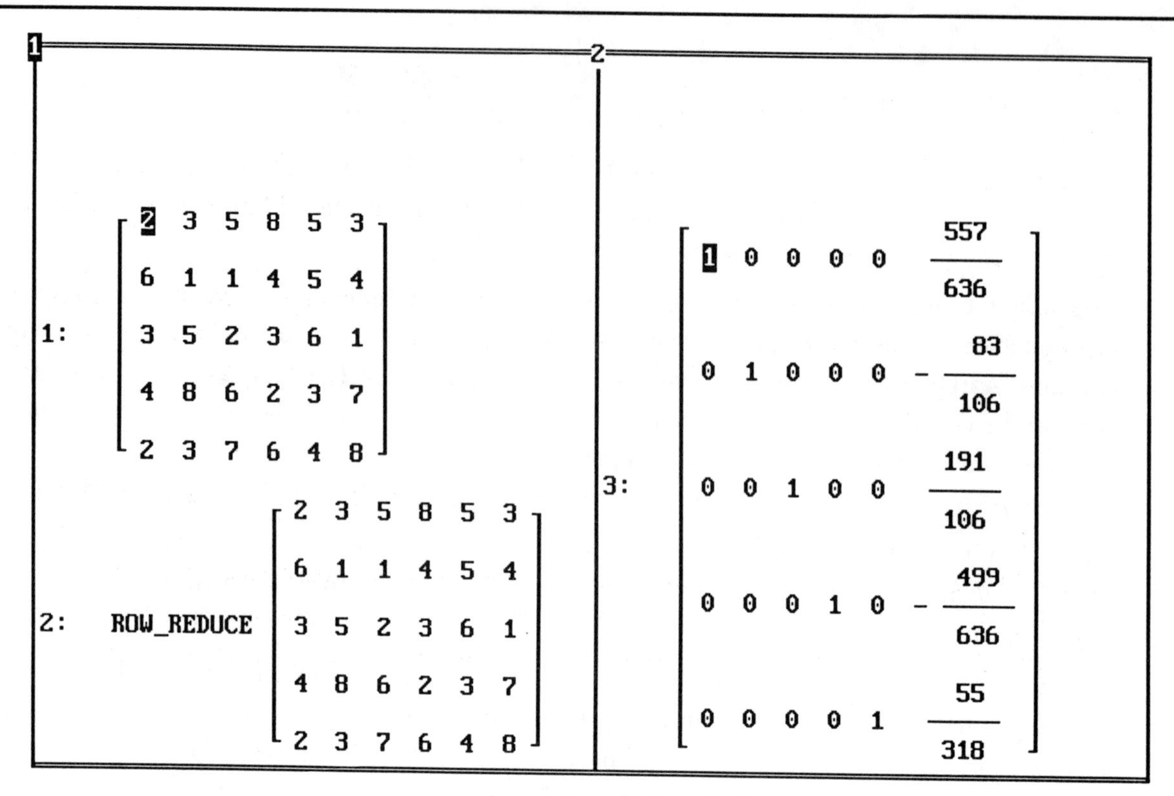

Figure 9.5: **A basis for R^5**

Solution: There are several correct approaches to this problem but we will take the most efficient one. Using **Declare Matrix** enter the 5 by 6 matrix $\begin{bmatrix} 2 & 3 & 5 & 8 & 5 & 3 \\ 6 & 1 & 1 & 4 & 5 & 4 \\ 3 & 5 & 2 & 3 & 6 & 1 \\ 4 & 8 & 6 & 2 & 3 & 7 \\ 2 & 3 & 7 & 6 & 4 & 8 \end{bmatrix}$ whose columns are the six given vectors.

DERIVE Hint: You may find it easier to enter the given vectors as *rows* in a 6 by 5 matrix and then take its transpose. If you find a mistake after it's on the screen, use F3 to bring it back to the author line for editing.

Now, ROW_REDUCE this matrix as in expressions 2 and 3 of Figure 9.5. This solves both problems at once. Since the first five columns reduce to the 5 by 5 identity matrix, S is a basis for R^5 and we see the desired coordinate vector relative to the ordered basis S as the last column in expression 3: $\left(\frac{557}{636}, -\frac{83}{106}, \frac{191}{106}, -\frac{499}{636}, \frac{55}{318}\right)$.

Solved Problem 5: Show that $S = \{2 - x + x^3,\ x + 3x^2,\ 5 - x^2 - x^3,\ 2 + x + 2x^2 + 4x^3\}$ is a basis for the space P_3 of polynomials of degree no more than 3. Find the coordinate vector of $1 + x + x^2 + x^3$ relative to the ordered basis S.

Solution: We will make use of the theorem that n linearly independent vectors in a vector space of dimension n is a basis. Since P_3 has dimension 4, we need only show that the four polynomials are linearly independent. **Author** each of the following as a separate expression.

```
P(x):= 2-x+x^3     Q(x):= x+3x^2     R(x):= 5-x^2-x^3     S(x):= 2+x+2x^2+4x^3
```

Next **Author** the expression `aP(x)+bQ(x)+cR(x)+dS(x)` as in expression 5 of Figure 9.6. **Simplify** expression 5. From expression 6 we see that $aP(x) + bQ(x) + cR(x) + dS(x) = 0$ if and only if a, b, c, and d are solutions of the following system of equations.

$$\begin{aligned} a - c + 4d &= 0 \\ 3b - c + 2d &= 0 \\ -a + b + d &= 0 \\ 2a + 5c + 2d &= 0 \end{aligned}$$

This system appears as expression 7 of Figure 9.6. If we **soLve** , we get expression 8, which shows that $a = b = c = d = 0$ is the unique solution. Thus the four functions are linearly independent.

We could also use the other technique described in Solved Problem 3. If $aP(x) + bQ(x) + cR(x) + dS(x)$ is the zero function, then it must be 0 for *all* values of x. Use **Manage Substitute** to substitute the values 0, 1, 2, and 3 for x in expression 5 of Figure 9.6 then **soLve** the resulting system. Although this method is often more laborious to execute by pencil and paper than the first, the reverse is true with *DERIVE*.

To find the coordinates of $1+x+x^2+x^3$ relative to the ordered basis S, set its coefficients equal to those of $aP(x) + bQ(x) + cR(x) + dS(x)$ as seen in expression 9 and then **soLve.**

We read the coordinate vector as $\left(-\frac{36}{79}, \frac{10}{79}, \frac{17}{79}, \frac{33}{79}\right)$. We have checked our work in expressions 11 and 12.

4: $S(x) := 2 + x + 2x^2 + 4x^3$

5: $a P(x) + b Q(x) + c R(x) + d S(x) = 0$

6: $x^3 (a - c + 4d) + x^2 (3b - c + 2d) - x(a - b - d) + 2a + 5c + 2d = 0$

7: $[a - c + 4d, 3b - c + 2d, a - b - d, 2a + 5c + 2d]$

8: $[a = 0, b = 0, c = 0, d = 0]$

9: $[a - c + 4d = 1, 3b - c + 2d = 1, -(a - b - d) = 1, 2a + 5c + 2d = 1$

10: $\left[a = -\frac{36}{79}, b = \frac{10}{79}, c = \frac{17}{79}, d = \frac{33}{79}\right]$

11: $\left[-\frac{36}{79}\right] P(x) + \frac{10}{79} Q(x) + \frac{17}{79} R(x) + \frac{33}{79} S(x)$

12: $x^3 + x^2 + x + 1$

Figure 9.6: A basis for P_3

Exercises

1. Determine if the following sets are linearly independent. If they are linearly dependent write one element as a linear combination of the others. Unless otherwise instructed by your teacher, you may use any method presented in this chapter.

   (a) $(2,5,6,8,3,1)$, $(5,3,9,1,4,2)$, $(4,4,1,2,7,4)$, $(8,9,4,1,3,5)$

   (b) $(4,2,3,4,6,7)$, $(3,2,4,1,3,2)$, $(2,1,1,3,1,2)$, $(5,3,6,2,8,7)$

   (c) $\begin{bmatrix} 1 & 3 & 2 \\ 3 & 1 & 4 \\ 2 & 8 & 7 \end{bmatrix}, \begin{bmatrix} 3 & 7 & 6 \\ 2 & 5 & 4 \\ 8 & 1 & 2 \end{bmatrix}, \begin{bmatrix} 3 & 6 & 9 \\ 2 & 3 & 1 \\ 5 & 0 & 1 \end{bmatrix}, \begin{bmatrix} 5 & 4 & 8 \\ 6 & 1 & 3 \\ 4 & 4 & 2 \end{bmatrix} \begin{bmatrix} 3 & 8 & 5 \\ 1 & 2 & 3 \\ 9 & 9 & 2 \end{bmatrix}$

   (d) $\cos x, \sin x, \cos(2x), \sin(2x)$

   (e) $\cos^2 x, \cos(2x), 8, e^x$

   (f) $(x-r)^2, (x-s)^2, (x-t)^2$, where r, s, and t are distinct numbers.

   (g) $\sin(x-1)$, $\sin(x-2)$, $\sin(x-3)$

   (h) $\sin x$, $\sin(x-\pi)$, $\sin(x-\frac{\pi}{2})$

   (i) $\ln(x^6+10x^4+33x^2+36)$, $\ln(x^6+9x^4+27x^2+27)$, $\ln(x^6+11x^4+40x^2+48)$

2. Find a basis for the subspace spanned by the given elements in exercise 1.

3. In each of the following cases, show that the set S is a basis for the given vector space and find the coordinates of the given element $\mathbf{v}$ relative to S as an ordered basis.

   (a) R^5; $S = \{(3,1,3,2,6,4), (4,5,7,2,4,3), (3,2,1,5,4), (2,9,1,4,4), (3,3,6,6,7)\}$;
   $\mathbf{v} = (2,4,1,2,3)$

   (b) P_3; $S = \{x^3-x+1,\ x^3+x^2+3,\ 2x^3+3x^2-x+4,\ x^3+4x^2+5x-2\}$;
   $\mathbf{v} = 3x^3-4x^2+2x-7$

   (c) $M_{2,2}$; $S = \left\{\begin{bmatrix} 2 & 7 \\ 3 & 9 \end{bmatrix}, \begin{bmatrix} 3 & 5 \\ 2 & 5 \end{bmatrix}, \begin{bmatrix} 5 & 2 \\ 9 & 7 \end{bmatrix}, \begin{bmatrix} 3 & 6 \\ 1 & 4 \end{bmatrix}\right\}$; $\mathbf{v} = \begin{bmatrix} 4 & 2 \\ 9 & 0 \end{bmatrix}$

4. In Solved Problem 5 show directly that the four polynomials span P_3 by showing the equation $f + gx + hx^2 + ix^3 = ap(x) + bq(x) + cr(x) + ds(x)$ has a solution for a, b, c, and d, no matter what the values of f, g, h, and i are.

5. Suppose $S = \{\mathbf{v_1}, \mathbf{v_2}, \mathbf{v_3}, \mathbf{v_4}\}$ is an ordered basis for R^4 and that the coordinate vectors relative to S of $(2,3,5,1)$ $(3,7,2,4)$ $(5,1,1,3)$ and $(3,1,3,3)$ are respectively, $(1,5,2,3)$ $(4,1,2,2)$ $(2,2,7,6)$ and $(4,4,1,2)$ Find $\mathbf{v_1}, \mathbf{v_2}, \mathbf{v_3}$, and $\mathbf{v_4}$.

Exploration and Discovery

1. In Solved Problem 3 we concluded that the four functions e^x, $\sin x$, $\cos x$, and $\ln x$ are linearly independent by substituting the values 0, 1, 2, and 3 for x in the equation $ae^x + b\sin x + c\cos x + d\ln x = 0$ and solving the resulting system. In Solved Problem 5 we remarked that you could do the same for the four polynomials presented there.

   (a) Carry out the method of substituting the values 0, 1, 2, and 3 for x in the equation $aP(x) + bQ(x) + cR(x) + dS(x) = 0$ in Solved Problem 5 and solving the resulting system of equations.

   (b) At the end of Solved Problem 3 we asked what you would conclude if the system had nontrivial solutions. Let's try out our method on the three functions: $x + 1$, $x^3 - 3x^2 + 3x + 1$, and $2 - \cos(\frac{\pi x}{2})$. Substitute the values 0, 1, and 2 for x in the equation $a(x+1) + b(x^3 - 3x^2 + 3x + 1) + c\left(2 - \cos(\frac{\pi x}{2})\right) = 0$ and **soLve** the resulting system. Explain what happens.

   (c) Instead of substituting the values 0, 1, and 2 for x in the equation $a(x+1) + b(x^3 - 3x^2 + 3x + 1) + c\left(2 - \cos(\frac{\pi x}{2})\right) = 0$, try substituting 1, 2 and 3 for x. Explain what happens.

   (d) Try substituting 3, 4 and 5 for x. Explain what happens. Are $x+1, x^3 - 3x^2 + 3x + 1$, and $2 - \cos(\frac{\pi x}{2})$ linearly independent? Summarize what you have concluded about this method.

2. **For students who have studied calculus.** There is a method for establishing independence of functions on an interval that arises in differential equations. It uses the *Wronskian.* We denote the ith derivative of f by $f^{(i)}$. If $f_1, f_2, \ldots, f_n$ are functions on (a, b) that have derivatives of order $n - 1$, their Wronskian is defined as

$$W(f_1, f_2, \cdots, f_n) = \text{DET} \begin{bmatrix} f_1 & f_2 & \cdots & f_n \\ f'_1 & f'_2 & \cdots & f'_n \\ \cdot & \cdot & \cdots & \cdot \\ \cdot & \cdot & \cdots & \cdot \\ \cdot & \cdot & \cdots & \cdot \\ f_1^{(n-1)} & f_2^{(n-1)} & \cdots & f_n^{(n-1)} \end{bmatrix}$$

__Theorem__: If $W(f_1, f_2, \cdots, f_n)$ is not identically zero, then $f_1, f_2, \ldots, f_n$ are linearly independent.

Define the Wronskian by **Authoring**

```
wron(z):=det(vector(dif(z,x,i),i,0,dimension(z)-1)).
```

(You may wish to save this definition in a file for later use.)

(a) The theorem stated above is not difficult to prove. Try it. Hint: Suppose that $f_1, f_2, \ldots, f_n$ are linearly dependent and show that their Wronskian must be zero.

(b) **Author** and **Simplify** `wron([`Alt e`^x,sin x,cos x,ln(x^2+1)])`. Compare your conclusion with the one obtained in Solved Problem 3.

(c) Use the Wronskian to test the independence of the functions in Solved Problem 5.

(d) Determine the value of the Wronskian for $\{1, x, x^2, \cdots, x^n\}$. (Hint: Calculate it for several values of n and look at the ratio of the nth calculation to the $(n-1)$st.)

(e) Determine the value of the Wronskian for $\{e^x, e^{2x}, \cdots, e^{nx}\}$. (Hint: Calculate it for several values of n and look at the ratio of the nth calculation to the $(n-1)$st.)

LABORATORY EXERCISE 9.1

Sets of Vectors

Name ______________________________ **Due Date** ____________________

Let $A = \begin{bmatrix} 3 & 5 & 9 & 8 \\ 2 & 3 & 1 & 7 \\ 4 & 1 & 2 & 9 \\ 6 & 3 & 4 & 1 \end{bmatrix}$. Let S denote the subspace of R^4 spanned by $(3,2,1,1)$, $(1,1,3,1)$, $(2,1,-2,0)$, and $(4,3,4,2)$.

1. Find a spanning set for the null space of A.

2. Find a basis for S.

3. Find all vectors that are in both S and the null space of A.

4. The set in Part 3 above is a subspace of R^4. Find a basis for it.

CHAPTER 10

ROW SPACE, COLUMN SPACE, AND NULL SPACE

LINEAR ALGEBRA CONCEPTS

- **Row space**
- **Column space**
- **Null space**
- **Rank**
- **Nullity**

Introduction

In this chapter we look at methods for producing bases for important vector spaces associated with matrices.

Solved Problems

Solved Problem 1: Find bases for (a) the row space, (b) the column space and (c) the null space of the following matrix.

$$A = \begin{bmatrix} 1 & 5 & 2 & 4 & 4 & 7 \\ 3 & 2 & 4 & 9 & 1 & 3 \\ 5 & 2 & 4 & 8 & 5 & 7 \\ 9 & 9 & 10 & 21 & 10 & 17 \end{bmatrix}$$

Solution to (a): Enter the 4 by 6 matrix and name it A. We use the fact that *the nonzero rows of the reduced echelon form of A form a basis for the row space of A.* Thus we **Author** and **Simplify** `row_reduce(A)`. From expression 4 of Figure 10.1, we read a basis for the row space of A as

$$\left\{(1,0,0,-\frac{1}{2},2,2),(0,1,0,-\frac{3}{16},\frac{9}{8},\frac{13}{8}),(0,0,1,\frac{87}{32},-\frac{29}{16},-\frac{25}{16})\right\}$$

```
1
2:  a := [ 1  5   2   4   4   7 ]
         [ 3  2   4   9   1   3 ]
         [ 5  2   4   8   5   7 ]
         [ 9  9  10  21  10  17 ]

3:  ROW_REDUCE (a)

4:  [ 1  0  0   - 1/2      2       2    ]
    [ 0  1  0   - 3/16    9/8    13/8   ]
    [ 0  0  1    87/32  - 29/16  - 25/16 ]
    [ 0  0  0     0        0       0    ]

2
5:  ROW_REDUCE (a`)

6:  [ 1  0  0  1 ]
    [ 0  1  0  1 ]
    [ 0  0  1  1 ]
    [ 0  0  0  0 ]
    [ 0  0  0  0 ]
    [ 0  0  0  0 ]
```

Figure 10.1: **Row and column spaces**

Solution to (b): The column space of A is the row space of the transpose of A. Therefore, we may **Author** `row_reduce(A')` and **Simplify**. From expression 6 of Figure 10.1 we obtain the basis {(1,0,0,1), (0,1,0,1), (0,0,1,1)}. Notice as well that the dimensions of the row and column spaces are the same. (This is always true, and there should be a theorem to that effect in your text.)

There is an alternative method for finding a basis for the column space that does not require the row reduction of the transpose of A: *The columns of A corresponding to columns of the reduced form of A that contain leading "1s" form a basis for the column space.* From expression 4 of Figure 10.1 the first three columns of the reduced form of A contain leading "1s". Thus the first three columns of A; $\begin{bmatrix} 1 \\ 3 \\ 5 \\ 9 \end{bmatrix}, \begin{bmatrix} 5 \\ 2 \\ 2 \\ 9 \end{bmatrix}, \begin{bmatrix} 2 \\ 4 \\ 4 \\ 10 \end{bmatrix}$, also form a basis for the column space of A.

We emphasize that both {(1,0,0,1), (0,1,0,1), (0,0,1,1)} and {(1,3,5,9), (5,2,2,9), (2,4,4,10)} are correct answers. It is interesting to notice that these two methods produce entirely different bases, but the subspaces spanned by the two bases are the same.

Solution to (c): The null space of A is the subspace of solutions of $A\mathbf{x} = 0$; thus we must solve this system of equations and describe the solution set. There are two ways to use *DERIVE* to assist us. We will present both.

Method 1 for (c). Using expression 4 of Figure 10.1, we recover the solution ***with pencil and paper*** from the row reduced form of A.

$$\begin{aligned} x_4 &= r \\ x_5 &= s \\ x_6 &= t \\ x_1 &= \frac{1}{2}r - 2s - 2t \\ x_2 &= \frac{3}{16}r - \frac{9}{8}s - \frac{13}{8}t \\ x_3 &= -\frac{87}{32}r + \frac{29}{16}s + \frac{25}{16}t \end{aligned}$$

To parlay this into a basis for the null space, set each parameter equal to 1 and the others to 0 in turn, as illustrated in the following table. The three column vectors form the desired basis.

$r = 1\ s = 0\ t = 0$	$r = 0\ s = 1\ t = 0$	$r = 0\ s = 0\ t = 1$
$x_1 = \frac{1}{2}$	$x_1 = -2$	$x_1 = -2$
$x_2 = \frac{3}{16}$	$x_2 = -\frac{9}{8}$	$x_2 = -\frac{13}{8}$
$x_3 = -\frac{87}{32}$	$x_3 = \frac{29}{16}$	$x_3 = \frac{25}{16}$
$x_4 = 1$	$x_4 = 0$	$x_4 = 0$
$x_5 = 0$	$x_5 = 1$	$x_5 = 0$
$x_6 = 0$	$x_6 = 0$	$x_6 = 1$

$$\begin{bmatrix} \frac{1}{2} \\ \frac{3}{16} \\ -\frac{87}{32} \\ 1 \\ 0 \\ 0 \end{bmatrix} \qquad \begin{bmatrix} -2 \\ -\frac{9}{8} \\ \frac{29}{16} \\ 0 \\ 1 \\ 0 \end{bmatrix} \qquad \begin{bmatrix} -2 \\ -\frac{13}{8} \\ \frac{25}{16} \\ 0 \\ 0 \\ 1 \end{bmatrix}$$

Method 2 for (c). The idea here is to obtain the solution of $A\mathbf{x} = 0$ with the variables displayed by *DERIVE*. **Author** and **Simplify** `A.[x,y,z,u,v,w]`. The result is partially displayed in expression 8 of Figure 10.2. We want the solution set of this system, so we **soLve** it. As we are prompted for the solve variables, we accept x, y, and z, pressing Enter after each, but we *do not accept the solve variable* u. When *DERIVE* offers it, Delete the u and press Enter. Expression 8 of Figure 10.2 is the result. (If we accept u as the fourth solve variable, *DERIVE* introduces an unnecessary arbitrary constant @1.)

```
7:   a · [x, y, z, u, v, w]

8:   [x + 5 y + 2 z + 4 u + 4 v + 7 w, 3 x + 2 y + 4 z + 9 u + v + 3 w, 5 x + 2

         u - 4 (v + w)         3 u - 2 (9 v + 13 w)         2 (29 v + 25 w) - 87
9:   [x = ─────────────, y = ────────────────────, z = ────────────────────
               2                      16                          32

      u - 4 (v + w)   3 u - 2 (9 v + 13 w)   2 (29 v + 25 w) - 87 u
10:  [─────────────, ────────────────────, ──────────────────────, u, v, w
            2                 16                      32
```

Figure 10.2: **Null space, second method: Part I**

Now, use F3 to bring expression 9 to the author line. Using Ctrl S and Ctrl D to move the cursor, insert at the end before the square bracket ",u,v,w" then delete "x =", "y =" and "z =." When we have finished we will see expression 10 of Figure 10.2.

Highlight expression 10 and use **Manage Substitute** to replace u by 1 and v and w by 0. The result is expression 11 of Figure 10.3, and when it is simplified we see expression 12. Repeat this process with $u = 0, v = 1, w = 0$ and $u = 0, v = 0, w = 1$ to get the other two vectors in expressions 14 and 16 of Figure 10.3.

Method 2 takes more effort than method 1, but it avoids errors in transcribing expressions from the computer screen to paper, and if the basis vectors are to be used later, it avoids having to re-enter them. This may also be the preferred method when working with larger matrices.

$$11: \left[\frac{1 - 4\ (0 + 0)}{2}, \frac{3\ 1 - 2\ (9\ 0 + 13\ 0)}{16}, \frac{2\ (29\ 0 + 25\ 0) - 87\ 1}{32}, 1, 0, 0\right]$$

$$12: \left[\frac{1}{2}, \frac{3}{16}, -\frac{87}{32}, 1, 0, 0\right]$$

$$13: \left[\frac{0 - 4\ (1 + 0)}{2}, \frac{3\ 0 - 2\ (9\ 1 + 13\ 0)}{16}, \frac{2\ (29\ 1 + 25\ 0) - 87\ 0}{32}, 0, 1, 0\right]$$

$$14: \left[-2, -\frac{9}{8}, \frac{29}{16}, 0, 1, 0\right]$$

$$15: \left[\frac{0 - 4\ (0 + 1)}{2}, \frac{3\ 0 - 2\ (9\ 0 + 13\ 1)}{16}, \frac{2\ (29\ 0 + 25\ 1) - 87\ 0}{32}, 0, 0, 1\right]$$

$$16: \left[-2, -\frac{13}{8}, \frac{25}{16}, 0, 0, 1\right]$$

Figure 10.3: **Null space, second method: Part II**

<u>Solved Problem 2</u>: Verify that *rank* + *nullity* = *number of columns* if $A = \begin{bmatrix} 3 & 5 & 1 & 7 \\ 2 & 8 & 6 & 3 \\ 1 & 5 & 2 & 1 \\ 4 & 8 & 5 & 9 \end{bmatrix}$.

<u>Solution</u>: Enter the 4 by 4 matrix using **Declare Matrix**; then **Author** and **Simplify** `row_reduce(#1)`. From expression 3 of Figure 10.4 we see that $(1, 0, 0, \frac{91}{30})$, $(0, 1, 0, -\frac{13}{30})$, and $(0, 0, 1, \frac{1}{15})$ form a basis for the row space. Thus the rank of A is 3.

Following the procedure outlined in Solved Problem 1, we obtain the single vector $\begin{bmatrix} -\frac{91}{30} \\ \frac{13}{30} \\ \frac{1}{15} \\ 1 \end{bmatrix}$ as a basis for the null space. Thus the nullity is 1, and *rank* + *nullity* = 3 + 1 = 4 = *number of columns*.

The reason that this formula works in general is clear from this example: The rank of a

$$
\text{2:}\quad \texttt{ROW\_REDUCE}\begin{bmatrix} 3 & 5 & 1 & 7 \\ 2 & 8 & 6 & 3 \\ 1 & 5 & 2 & 1 \\ 4 & 8 & 5 & 9 \end{bmatrix}
$$

$$
\text{3:}\quad \begin{bmatrix} 1 & 0 & 0 & \frac{91}{30} \\ 0 & 1 & 0 & -\frac{13}{30} \\ 0 & 0 & 1 & \frac{1}{15} \\ 0 & 0 & 0 & 0 \end{bmatrix}
$$

Figure 10.4: **Rank and nullity of a matrix**

matrix is the number of nonzero rows in the reduced form of A (which is the same as the number of columns containing leading "1s") and the nullity is the number of parameters in the solution of $A\mathbf{x} = 0$, which is the number of columns in the reduced form of A that do *not* contain leading "1s."

Exercises

1. For each of the following matrices, find bases for the row space, column space, and null space. In each case verify that *rank* + *nullity* = *number of columns*.

(a) $\begin{bmatrix} 2 & 3 & 8 & 4 & 7 \\ 5 & 1 & 3 & 9 & 6 \\ 2 & 4 & 8 & 5 & 2 \\ 5 & 0 & 3 & 8 & 11 \end{bmatrix}$

(b) $\begin{bmatrix} 2 & 3 & 4 & 7 & 6 & 2 \\ 2 & 3 & 4 & 3 & 2 & 1 \\ 2 & 3 & 4 & 1 & 9 & 7 \\ 4 & 6 & 8 & 0 & 1 & 3 \\ 4 & 6 & 8 & 4 & 2 & 4 \end{bmatrix}$

(c) $\begin{bmatrix} 5 & 1 & 1 & 8 & 2 & 3 & 5 \\ 10 & 2 & 2 & 4 & 1 & 5 & 4 \\ 5 & 1 & 1 & 9 & 1 & 1 & 5 \\ 10 & 2 & 2 & 4 & 5 & 6 & 3 \end{bmatrix}$

2. Find a basis for the subspace of R^5 spanned by $(2,4,3,6,1)$, $(3,1,5,7,2)$, $(5,5,8,13,3)$, $(-1,3,-2,-1,-1)$, $(8,6,13,20,5)$. (Hint: This subspace is the row space of some matrix.)

3. Find a basis for the subspace of R^6 each element of which is orthogonal to all three of the vectors $(1,2,4,6,7,2), (4,1,7,9,8,6), (4,1,5,2,3,4)$. (Hint: This subspace is the null space of some matrix.)

4. Find the intersection of the null spaces of the matrices $\begin{bmatrix} 4 & 5 & 9 & 2 & 3 & 4 \\ 3 & 7 & 6 & 5 & 2 & 1 \end{bmatrix}$ and $\begin{bmatrix} 2 & 6 & 3 & 5 & 9 & 2 \\ 1 & 2 & 1 & 2 & 3 & 4 \\ 3 & 1 & 2 & 6 & 9 & 1 \end{bmatrix}$. (Hint: Is the intersection the null space of a single matrix?)

Exploration and Discovery

Let S be the subspace of R^5 spanned by $(3,1,5,7,2)$ and $(5,5,8,13,3)$.

1. Find a 4 by 5 matrix whose null space is S. Explain your method.

2. Find a 3 by 5 matrix whose null space is S. Explain your method.

3. Find a 2 by 5 matrix whose null space is S. Explain the difficulty encountered here.

4. Suppose you are given a subspace S of R^n of dimension k. If there is an m by n matrix whose null space is S, what can you say about m in general? Explain.

LABORATORY EXERCISE 10.1

Subspaces Associated with a Matrix

Name ______________________________ Due Date ________________

Let $A = \begin{bmatrix} 7 & 0 & 1 & 8 & 2 & 3 & 5 \\ 5 & 1 & 2 & 4 & 1 & 0 & 4 \\ 3 & 3 & 1 & 9 & 1 & 2 & 1 \\ 9 & -2 & 2 & 3 & 2 & 1 & 8 \end{bmatrix}$

1. Find a basis for the null space of A.

2. Multiply each basis vector above by A. Explain the meaning of your answers.

3. Find a basis for the row space of A.

4. Find a basis for the column space of A by row-reducing the transpose of A.

5. Find a basis for the column space of A by row-reducing A itself.

CHAPTER 11

INNER PRODUCT SPACES

LINEAR ALGEBRA CONCEPTS
• **General inner products**

Introduction

If A is a given invertible n by n matrix, and $\mathbf{x}$ and $\mathbf{y}$ are vectors in R^n, we will denote $(A\mathbf{x}) \cdot (A\mathbf{y})$ by $<\mathbf{x}, \mathbf{y}>$. This is called the *inner product on* R^n *generated by* A; it has the familiar properties of the dot product. (Indeed, if A is the identity matrix, then $<\mathbf{x}, \mathbf{y}> = \mathbf{x} \cdot \mathbf{y}$.) Some texts may define $<\mathbf{x}, \mathbf{y}>$ by $(A\mathbf{x})^t(A\mathbf{y})$, which amounts to the same thing; however, since *DERIVE* will not accept this syntax, we will use $(A\mathbf{x}) \cdot (A\mathbf{y})$, which is entered in *DERIVE* as `(A.u).(A.v)` . (Watch the periods.)

Solved Problems

<u>Solved Problem 1</u>: Let $<\mathbf{x}, \mathbf{y}>$ denote the inner product on R^4 generated by the matrix

$$A = \begin{bmatrix} 2 & 3 & 8 & 1 \\ 3 & 1 & 5 & 2 \\ 3 & 3 & 9 & 7 \\ 2 & 1 & 4 & 1 \end{bmatrix}.$$ Let $\mathbf{u} = (2, 4, 3, 1)$ and $\mathbf{v} = (3, 6, 1, 4)$.

1. Calculate $<\mathbf{u}, \mathbf{v}>$.
2. Calculate the length, $||\mathbf{u}||$, of $\mathbf{u}$ with respect to this inner product.
3. Find a nonzero vector that is orthogonal to $\mathbf{u}$ with respect to this inner product.

<u>Solution</u>: <u>Part 1</u>. Use **Declare Matrix** to enter the matrix and name it A. Next **Author** `u:=[2,4,3,1]` and `v:=[3,6,1,4]`. Finally, **Author** and **Simplify** `(A.u).(A.v)`. (As we can see in expression 5 of Figure 11.1, *DERIVE* discards the first pair of parentheses.) The inner product appears in expression 6 of Figure 11.1.

<u>Part 2</u>. Recall that $||\mathbf{u}|| = \sqrt{<\mathbf{u}, \mathbf{u}>}$. Thus we **Author** and **Simplify** Alt Q `((A.u).(A.u))`. The length appears in expression 8 of Figure 11.1.

```
2:   a := [ 2  3  8  1 ]
          [ 3  1  5  2 ]
          [ 3  3  9  7 ]
          [ 2  1  4  1 ]

3:   u := [2, 4, 3, 1]

4:   v := [3, 6, 1, 4]

5:   a · u · (a · v)

6:   5980

7:   √(a · u · (a · u))

8:   √5555

9:   a · u · (a · [p, q, r, s])

10:  361 p + 327 q + 1015 r + 480 s
```

Figure 11.1: **An inner product on R^4 generated by a matrix**

Part 3. We want to find p, q, r, and s (not all zero) such that $<\mathbf{u}, (p,q,r,s)>= 0$. Thus, we **Author** and **Simplify** `(A.u).(A.[p,q,r,s])` as seen in expressions 9 and 10 of Figure 11.1. From expression 10 we see that there are infinitely many solutions, one of which is $p = 1, q = -\frac{361}{327}, r = 0, s = 0$, yielding the vector $(1, -\frac{361}{327}, 0, 0)$.

Solved Problem 2: **For students who have studied calculus.** If f and g are continuous on the closed interval $[a, b]$, we define the inner product

$$<f, g> \text{ to be } \int_a^b f(x)g(x)\,dx$$

In this problem we will take $[a, b] = [-1, 1]$.

Find $||\sin x||$ and show that if $k \neq \pm 1$ is an integer, then $\sin(\pi x)$ and $\sin(k\pi x)$ are orthogonal.

```
1:   SIN (x)^2

      1
2:   ∫    SIN (x)^2 dx
      -1

3:   1 - SIN (1) COS (1)

4:   √(1 - SIN (1) COS (1))

5:   0.738478

6:   SIN (π x) SIN (k π x)

      1
7:   ∫    SIN (π x) SIN (k π x) dx
      -1

          2 SIN (π k)
8:   ---------------------
     π (k + 1) (1 - k)
```

Figure 11.2: **An inner product on $C(-1, 1)$**

Solution: $||\sin x|| = \sqrt{< \sin x, \sin x >} = \sqrt{\int_{-1}^{1} \sin^2 x\, dx}$. **Author** `(sin x)^2`, and use **Calculus Integrate**. Set the lower limit to -1 and use the Tab key to set the upper limit to 1. **Simplify** to obtain the result displayed in expression 3 of Figure 11.2. **Author** Alt Q `#3` and **approX** to get $||\sin x|| = \sqrt{1 - \sin 1 \cos 1} \approx 0.738478$.

For the second part of the exercise, **Author** `sin(pi x)sin(k pi x)`. Use **Calculus Integrate**, set the limits on the integral to -1 and 1, and **Simplify** to obtain the results in expressions 7 and 8 of Figure 11.2. Since k is an integer and $k \neq \pm 1$, we see that this reduces to 0. Thus, the two functions are orthogonal.

Solved Problem 3: **For students who have studied calculus.** Using the inner product of Solved Problem 2, find a quadratic polynomial with leading coefficient 1 that is orthogonal to both $\sin x$ and $\cos x$

Solution: If we are going to calculate an inner product several times, it is convenient to define it as a function. **Author** `inp(f,g):=int(fg,x,-1,1)` as seen in Figure 11.3. Next

```
                      1
                     ⌠
2:    INP(f, g) :=   ⌡    f g dx
                     -1

          2
3:    y := x   + a x + b

4:    [INP(y, SIN(x)), INP(y, COS(x))]

5:    [2 a SIN(1) - 2 a COS(1), 4 COS(1) + (2 b - 2) SIN(1)]

6:    [a = 0, b = 1 - 2 COT(1)]
```

Figure 11.3: **A quadratic orthogonal to** $\sin x$ **and** $\cos x$

we name a second-degree polynomial y using `y:=x^2+ax+b`. We wish to find a and b so that both <y, $\sin x$> and <y, $\cos x$> are zero. Thus, we **Author**, **Simplify** and **soLve** `[inp(y,sin x),inp(y,cos x)]`. The result displayed in expression 5 of Figure 11.3 tells us that the required polynomial is $x^2 + 1 - 2\cot 1$.

Exercises

1. This problem refers to the inner product from Solved Problem 1. Be careful to use it and not the usual dot product. You may find it convenient to define the matrix A and then **Author** `inp(x,y):=(a.x).(a.y)` to define a function for the inner product.

   (a) Find $<(2,5,3,4), (7,1,4,2)>$.

   (b) Find the angle (in radians) between $(2,5,3,4)$ and $(7,1,4,2)$ for this inner product. Hint: The cosine of the angle between $\mathbf{u}$ and $\mathbf{v}$ is $\dfrac{<\mathbf{u},\mathbf{v}>}{||\mathbf{u}||\ ||\mathbf{v}||}$.

   (c) Find a unit vector in the direction of $(3,6,1,3)$.

   (d) Find all vectors simultaneously orthogonal to $(2,1,5,3)$, $(3,1,2,2)$ and $(2,4,3,5)$.

2. **For students who have studied calculus.** Let $<f, g> = \int_{-1}^{1} f(x)g(x)\,dx$.

   (a) Show that if p and q are positive integers and $p \neq q$, then $||\cos(p\pi x)|| = 1$ and that $\cos(p\pi x)$ is orthogonal to $\cos(q\pi x)$.

   (b) Find a polynomial in P_2 of length 1 that is orthogonal to both e^x and e^{-x}.

3. Let $\mathbf{a} = (1,2,3,2), \mathbf{b} = (1,3,2,4)$ and $\mathbf{c} = (3,2,2,1)$. Suppose that $<\mathbf{u}, \mathbf{v}>$ is an inner product on R^4 and that the following are known.

   $<\mathbf{a}, \mathbf{a}> = 2$ $\quad <\mathbf{a}, \mathbf{b}> = 3$ $\quad <\mathbf{a}, \mathbf{c}> = 1$ $\quad <\mathbf{b}, \mathbf{b}> = 1$ $\quad <\mathbf{b}, \mathbf{c}> = 4$ $\quad <\mathbf{c}, \mathbf{c}> = 2$

   Find $<(-1,3,3,5), (3,1,3,-1)>$. (Hint: Can the given vectors be written as linear combinations of $\mathbf{a}, \mathbf{b}$ and $\mathbf{c}$?)

4. **For students who have studied calculus.** Compute the inner product $\int_{-1}^{1}(ax+b)(cx+d)\,dx$. Find a matrix A such that the inner product $<(a,b), (c,d)>$ that it generates on R^2 equals $\int_{-1}^{1}(ax+b)(cx+d)\,dx$. Hint: Assume A is diagonal: $\begin{bmatrix} r & 0 \\ 0 & s \end{bmatrix}$ (For more details, see Exploration and Discovery Problem 2.)

Exploration and Discovery

1. Let A be an n by n matrix and let $<\mathbf{u}, \mathbf{v}>= (A\mathbf{x}) \cdot (A\mathbf{y})$ be the inner product on R^n generated by A. The first three are not computer problems.

   (a) Show that $S = A^tA$ is symmetric; that is $S^t = S$.

   (b) Show that $<\mathbf{x}, \mathbf{y}>= \mathbf{x}^tS\mathbf{y}$. (Recall that $(A\mathbf{x}) \cdot (A\mathbf{y}) = (A\mathbf{x})^t(A\mathbf{y})$.)

   (c) In (b) above we see that the inner product can be expressed as $<\mathbf{x}, \mathbf{y}>= \mathbf{x}^tS\mathbf{y}$, where S is symmetric. Let $S = \begin{bmatrix} 1 & 2 \\ 2 & 1 \end{bmatrix}$ and $\mathbf{x} = \begin{bmatrix} 1 \\ -1 \end{bmatrix}$. Find $\mathbf{x}^tS\mathbf{x}$. Discuss the conclusions you draw from this.

   (d) Let $\mathbf{u} = (1, 2)$ and $\mathbf{v} = (3, 4)$. Find a matrix A such that $<\mathbf{u}, \mathbf{v}> =0$.

   (e) Let $\mathbf{u} = (1, 2)$ and $\mathbf{v} = (2, -1)$. Find a matrix A, other than the identity, such that $<\mathbf{u}, \mathbf{v}> =0$.

   (f) Describe all 2 by 2 matrices A that "preserve orthogonality" that is, if $\mathbf{u}.\mathbf{v} = 0$ then $<\mathbf{u}, \mathbf{v}> =0$. (Hint: Consider the standard basis vectors.)

   (g) Describe all n by n matrices A that "preserve orthogonality."

2. **For students who have studied calculus.** Do Exercise 4.

   (a) Compute the inner product $\int_{-1}^{1} (ax^2 + bx + c)(dx^2 + ex + f)\, dx$.

   (b) Show that there is no diagonal matrix $\begin{bmatrix} r & 0 & 0 \\ 0 & s & 0 \\ 0 & 0 & t \end{bmatrix}$ such that the inner product $<(a, b, c), (d, e, f)>$ that it generates on R^3 equals $\int_{-1}^{1} (ax^2 + bx + c)(dx^2 + ex + f)\, dx$. Hint: Assume there is such a matrix and arrive at a contradiction.

   (c) Find a matrix A such that the inner product $<(a, b, c), (d, e, f)>$ that it generates on R^3 equals $\int_{-1}^{1} (ax^2 + bx + c)(dx^2 + ex + f)\, dx$. Hint: Try assuming A has the form $\begin{bmatrix} r & 0 & 0 \\ 0 & s & 0 \\ u & 0 & t \end{bmatrix}$.

   (d) Explore the situation for P_3 and R^4.

LABORATORY EXERCISE 11.1

Inner Products

Name ______________________ **Due Date** ______________

Let <**x**, **y**> denote the inner product on R^3 generated by $\begin{bmatrix} 2 & 3 & 1 \\ 1 & 5 & 2 \\ 3 & 1 & 1 \end{bmatrix}$. Let $\mathbf{u} = (1, 3, 1)$ and $\mathbf{v} = (2, 1, 1)$.

1. Calculate <**u**, **v**>.

2. Calculate the lengths of **u** and **v** with respect to this inner product.

3. Find a nonzero vector that is orthogonal to **u** with respect to this inner product.

4. Find a nonzero vector that is simultaneously orthogonal to **u** and **v** with respect to this inner product.

CHAPTER 12

ORTHONORMAL BASES AND THE GRAM-SCHMIDT PROCESS

LINEAR ALGEBRA CONCEPTS

- **Orthonormal basis**
- **Gram-Schmidt process**

Introduction

An orthonormal basis for an inner product space behaves very much like the standard basis for R^n. In particular, coordinate vectors are easy to compute in terms of the inner product. The Gram-Schmidt process is a method by which we may construct such a basis for a finite dimensional inner product space. It is often tedious to carry out with pencil and paper, but *DERIVE* makes it easy.

Solved Problems

<u>Solved Problem 1</u>: Verify that

$$S = \left\{\left(\tfrac{1}{\sqrt{7}},\tfrac{2}{\sqrt{7}},\tfrac{1}{\sqrt{7}},\tfrac{1}{\sqrt{7}}\right),\left(-\tfrac{1}{\sqrt{42}},-\tfrac{2}{\sqrt{42}},-\tfrac{1}{\sqrt{42}},\tfrac{6}{\sqrt{42}}\right),\left(-\tfrac{1}{\sqrt{30}},-\tfrac{2}{\sqrt{30}},\tfrac{5}{\sqrt{30}},0\right),\left(\tfrac{2}{\sqrt{5}},-\tfrac{1}{\sqrt{5}},0,0\right)\right\}$$

is an orthonormal basis for R^4 and find $[(1,2,3,4)]_S$ (the coordinate vector of $(1,2,3,4)$ relative to the ordered basis S).

<u>Solution</u>: **Author** and name the four vectors as follows. (See Figure 12.1.)

```
x:=[1,2,1,1]/sqrt(7)            y:=[-1,-2,-1,6]/sqrt(42)
z:=[-1,-2,5,0]/sqrt(30)           w:=[2,-1,0,0]/sqrt(5)
```

The first thing that might occur to us is to check the lengths and all possible inner products of these four vectors one by one. *DERIVE* can help us do that, but it is very laborious and there is a much more efficient way. We will enter the matrix whose rows are the

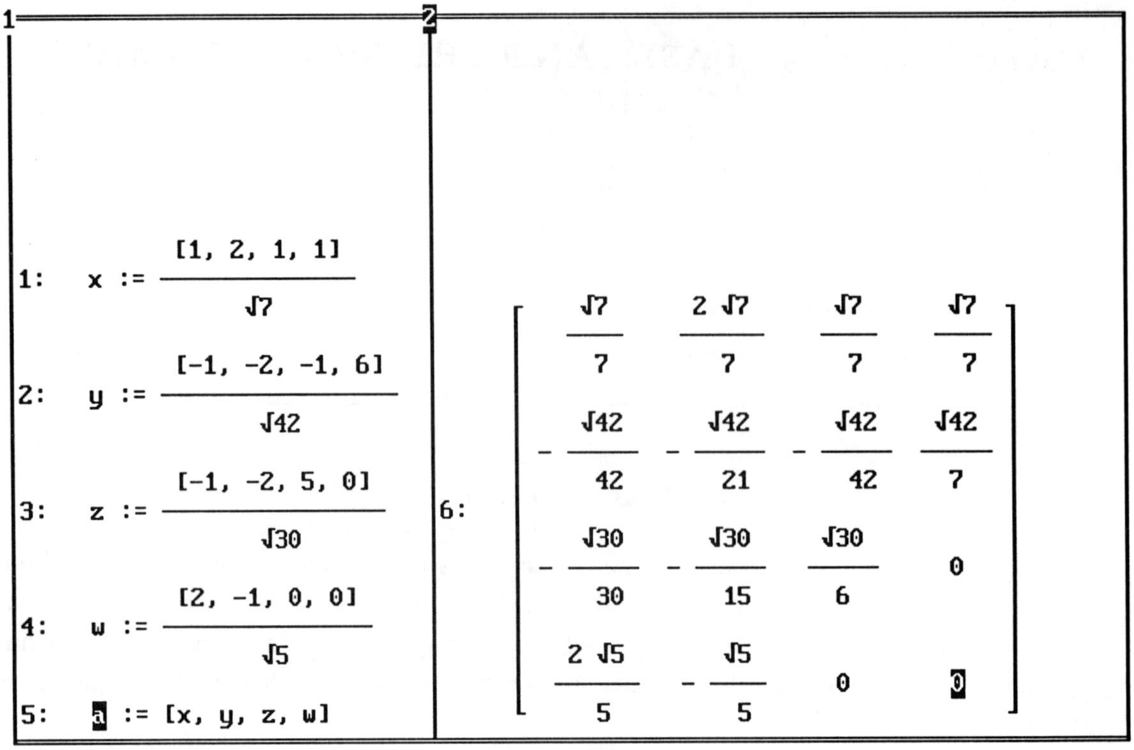

$$1: \quad x := \frac{[1,\ 2,\ 1,\ 1]}{\sqrt{7}}$$

$$2: \quad y := \frac{[-1,\ -2,\ -1,\ 6]}{\sqrt{42}}$$

$$3: \quad z := \frac{[-1,\ -2,\ 5,\ 0]}{\sqrt{30}}$$

$$4: \quad w := \frac{[2,\ -1,\ 0,\ 0]}{\sqrt{5}}$$

$$5: \quad a := [x,\ y,\ z,\ w]$$

$$6: \quad \begin{bmatrix} \frac{\sqrt{7}}{7} & \frac{2\sqrt{7}}{7} & \frac{\sqrt{7}}{7} & \frac{\sqrt{7}}{7} \\ -\frac{\sqrt{42}}{42} & -\frac{\sqrt{42}}{21} & -\frac{\sqrt{42}}{42} & \frac{\sqrt{42}}{7} \\ -\frac{\sqrt{30}}{30} & -\frac{\sqrt{30}}{15} & \frac{\sqrt{30}}{6} & 0 \\ \frac{2\sqrt{5}}{5} & -\frac{\sqrt{5}}{5} & 0 & 0 \end{bmatrix}$$

Figure 12.1: **An orthonormal basis for R^4**

four vectors and multiply it by its own transpose. **Author** and **Simplify** the expression `A:=[x,y,z,w]`. The result is displayed in expression 6 of Figure 12.1. To calculate the product of A with its transpose, **Author** and **Simplify** `A.A'`. We see in expression 8 of Figure 12.2 that the product is the identity matrix. (We leave it to the reader to explain why we may conclude that the four vectors form an orthonormal basis.)

DERIVE Hint: There are several alternatives for entering the matrix A into *DERIVE* other than the one described above. We could use **Declare Matrix** to enter the matrix in Figure 12.1, but it's very tedious to do and also very easy to make a mistake. The unsure typist may prefer the following method. Enter each row separately using **Declare vectoR** to enter the numerators; then divide the result by the appropriate square root. For example, to enter the first vector, use **Declare vectoR** to enter the numerators `[1,2,1,1]`. Next **Author** `#1/sqrt(7)`. Once the four vectors have been entered, assemble the matrix using **Declare vectoR** with dimension 4 and enter the expression number of each row vector.

```
7:   a · a`

     ┌ 1  0  0  0 ┐
     │ 0  1  0  0 │
8:   │ 0  0  1  0 │
     └ 0  0  0  1 ┘

9:   b := [1, 2, 3, 4]

10:  [b · x, b · y, b · z, b · w]

      ┌ 12 √7    8 √42    √30    ┐
11:   │ ─────,  ─────,   ───,  0 │
      └   7       21       3     ┘

12:  a · b

      ┌ 12 √7    8 √42    √30    ┐
13:   │ ─────,  ─────,   ───,  0 │
      └   7       21       3     ┘
```

Figure 12.2: **Checking an orthonormal basis for R^4**

For the second part of the problem we use the fact that the coordinate vector of **b** with respect to the ordered orthonormal basis $\{\mathbf{x}, \mathbf{y}, \mathbf{z}, \mathbf{w}\}$ is given by the following.

$$[\mathbf{b}]_S = (\mathbf{b} \cdot \mathbf{x}, \mathbf{b} \cdot \mathbf{y}, \mathbf{b} \cdot \mathbf{z}, \mathbf{b} \cdot \mathbf{w})$$

Author `b:=[1,2,3,4]`. Next **Author** and **Simplify** the expression `[b.x,b.y,b.z,b.w]`. The coordinate vector appears in expression 11 of Figure 12.2.

Notice that in expression 13 we have calculated the product, $A\mathbf{b}$, and that this also gives the proper coordinate vector. Why does this work?

<u>**Solved Problem 2**</u>: Using the inner product on $C(-1,1)$ defined by $<f,g>= \int_{-1}^{1} f(x)g(x)\,dx$, verify that $S = \left\{\frac{1}{\sqrt{2}}, \sin(\pi x), \sin(2\pi x), \sin(3\pi x)\right\}$ is an orthonormal set. Find the projection of $f(x) = x$ onto the subspace W spanned by these four vectors.

```
2:    b := SIN (π x)

3:    c := SIN (2 π x)

4:    d := SIN (3 π x)
```

$$5: \quad \text{INP}(f, g) := \int_{-1}^{1} f\, g\, dx$$

```
6:    INP (a, a)

7:    1

8:    INP (a, b)

9:    0

10:   INP (x, a) a + INP (x, b) b + INP (x, c) c + INP (x, d) d
```

$$11: \quad \frac{2 \text{ SIN}(3 \pi x)}{3 \pi} - \frac{\text{SIN}(2 \pi x)}{\pi} + \frac{2 \text{ SIN}(\pi x)}{\pi}$$

Figure 12.3: **The projection of x onto a subspace of $C(-1,1)$**

Solution: **Author** the following as separate expressions.

`a:=1/sqrt(2)`   `b:=sin(pi x)`   `c:=sin(2pi x)`   `d:=sin(3pi x)`

To assist in computing the inner product, **Author** `INP(f,g):=int(fg,x,-1,1)` as we did in Chapter 11. (See Figure 12.3.) To test for orthonormality we must calculate the inner product of all pairs of these elements. We have calculated INP(a,a) and INP(a,b) in expressions 6 through 9 of Figure 12.3. We leave it to the reader to complete the process.

Since S is orthonormal, the projection of **v** onto W is given by **<v,a>a+<v,b>b+<v,c>c+<v,d>d**. Thus we **Author** and **Simplify**

```
inp(x,a)a+inp(x,b)b+inp(x,c)c+inp(x,d)d
```

The result is displayed in expressions 10 and 11 of Figure 12.3.

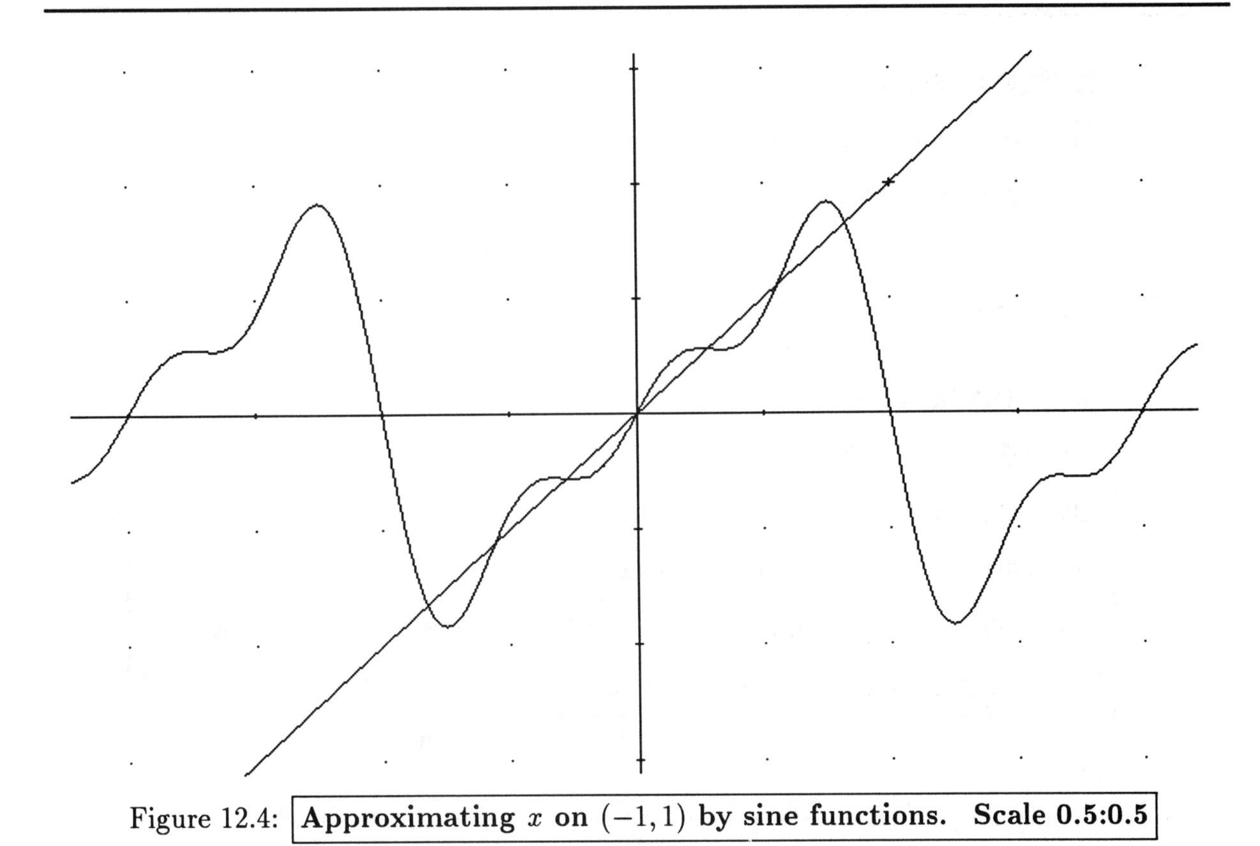

Figure 12.4: **Approximating x on $(-1,1)$ by sine functions. Scale 0.5:0.5**

Recall that the projection of a vector **v** onto a subspace W may be thought of as the best approximation of **v** by a vector in W, in the sense that among all elements **w** in W, $||\mathbf{v}-\mathbf{w}||$ is minimal when **w** is the projection of **v** onto W. The graph of x along with its projection onto W appears in Figure 12.4. The graph is made by highlighting expression 11 in Figure 12.3 and then using **Plot Overlay Plot**. To get the graph of $y = x$ on the same screen, use **Algebra** to get back to the algebra window, **Author x**, then **Plot** again.

It should be noted how the graphs appear close together on $(-1,1)$ until we approach the endpoints. (Our scale is `x:0.5 y:0.5`.) We do not expect the approximation to be good outside the interval $(-1,1)$, and this is clearly suggested by the graphs.

Solved Problem 3: Find an orthonormal basis for the subspace of R^5 spanned by the vectors $\mathbf{u_1} = (1,3,2,5,2), \mathbf{u_2} = (2,1,3,2,2), \mathbf{u_3} = (1,2,3,2,1)$.

Solution: The Gram-Schmidt process is tedious to carry out with pencil and paper, but *DERIVE* makes it easy. To avoid a lot of typing it is helpful to define a function for the projection: **Author** `P(u,v):=(u.v)/(v.v) v` for later use as in Figure 12.5.

4: $P(u, v) := \dfrac{u \cdot v}{v \cdot v} v$

5: $v1 := u1$

6: $v2 := u2 - P(u2, v1)$

7: $v3 := u3 - P(u3, v1) - P(u3, v2)$

8: $\left[\dfrac{v1}{|v1|}, \dfrac{v2}{|v2|}, \dfrac{v3}{|v3|}\right]$

$$
9: \quad \begin{bmatrix} \dfrac{\sqrt{43}}{43} & \dfrac{3\sqrt{43}}{43} & \dfrac{2\sqrt{43}}{43} & \dfrac{5\sqrt{43}}{43} & \dfrac{2\sqrt{43}}{43} \\ \dfrac{61\sqrt{13803}}{13803} & -\dfrac{32\sqrt{13803}}{13803} & \dfrac{79\sqrt{13803}}{13803} & -\dfrac{13\sqrt{13803}}{4601} & \dfrac{12\sqrt{13803}}{4601} \\ -\dfrac{23\sqrt{321}}{1284} & \dfrac{25\sqrt{321}}{856} & \dfrac{79\sqrt{321}}{2568} & -\dfrac{13\sqrt{321}}{856} & -\dfrac{71\sqrt{321}}{2568} \end{bmatrix}
$$

Figure 12.5: **Executing the Gram-Schmidt process in R^5**

Author the three vectors (not shown in Figure 12.5).

`u1:=[1,3,2,5,2]`, `u2:=[2,1,3,2,2]`, `u3:=[1,2,3,2,1]`

Next, **Author**

`v1:=u1` `v2:=u2-P(u2,v1)` `v3:=u3-P(u3,v1)-P(u3,v2)`

as separate expressions as shown in Figure 12.5. It is essential that vectors $\mathbf{v_1}$, $\mathbf{v_2}$, and $\mathbf{v_3}$ be defined in the order shown in Figure 12.5. (You may **Simplify** to see the intermediate steps if you wish, but we have not done so.)

The vectors $\mathbf{v_1}$, $\mathbf{v_2}$, $\mathbf{v_3}$ thus far defined form an *orthogonal* basis. We finish by normalizing each vector: **Author** `[v1/|v1|, v2/|v2|, v3/|v3|]` and **Simplify**. (Note: If you're using an earlier version of *DERIVE* , $|x|$ may not work. You'll have to define `LEN(x):=sqrt(x.x)`.)

The required orthonormal basis appears as the rows of the matrix displayed in expression 9 of Figure 12.5. We will not do so here, but you should use *DERIVE* to check this answer by showing that each of these are unit vectors and that they are mutually orthogonal. (Recall how we did this efficiently in Solved Problem 1.)

Solved Problem 4: Using the inner product on $C(-1,1)$ defined by $<f,g>= \int_{-1}^{1} f(x)g(x)\,dx$, find an orthonormal basis for the subspace spanned by $\{x, x^2, x^3\}$.

```
4:    INP (f, g) := ∫[-1,1] f g dx

5:    LEN (u) := √(INP (u, u))

                 INP (u, v)
6:    P (u, v) := ────────── v
                 INP (v, v)

7:    v1 := u1

8:    v2 := u2 - P (u2, v1)

9:    v3 := u3 - P (u3, v1) - P (u3, v2)

10:   [ v1/LEN (v1), v2/LEN (v2), v3/LEN (v3) ]
```

$$11: \quad \left[\frac{\sqrt{6}\,x}{2}, \frac{\sqrt{10}\,x^2}{2}, \frac{5\sqrt{14}\,x^3}{4} - \frac{3\sqrt{14}\,x}{4} \right]$$

Figure 12.6: **The Gram-Schmidt process in $C(-1,1)$**

Solution: This may appear to be more difficult than Solved Problem 3, but with *DERIVE* to do the calculations, the execution of the Gram-Schmidt process is the same in *any* inner product space. As before, to avoid a lot of typing, it is helpful to define a function for the projection: **Author** `P(u,v):=INP(u,v)/INP(v,v) v` for later use, where `INP(f,g):=int(fg,x,-1,1)` as in Solved Problem 2. (See Figure 12.6.)

After we **Author** and name the three vectors as `u1:=x, u2:=x^2, u3:=x^3` (not shown in Figure 12.6), we execute the Gram-Schmidt process exactly as in Solved Problem 1 using the projection function. **Author** `v1:=u1`, `v2:=u2-P(u2,v1)` and `v3=u3-P(u3,v1)-P(u3,v2)`.

(As before, you may **Simplify** to view the intermediate steps if you wish.) We can save a little typing by defining a length function for this inner product space. **Author** `len(f):=sqrt(INP(f,f))` as in expression 5 of Figure 12.6. Now we can change the orthogonal basis $\{\mathbf{v_1}, \mathbf{v_2}, \mathbf{v_3}\}$ to an orthonormal basis using
`[v1/len(v1), v2/len(v2), v3/len(v3)]`. **Simplify** to obtain the orthonormal basis displayed in expression 11 of Figure 12.6.

Exercises

1. Show that

$$S = \left\{ \left(\tfrac{1}{\sqrt{3}}, \tfrac{1}{\sqrt{3}}, 0, \tfrac{1}{\sqrt{3}} \right), \left(-\tfrac{1}{\sqrt{15}}, -\tfrac{1}{\sqrt{15}}, \tfrac{1}{\sqrt{15}}, \tfrac{2}{\sqrt{15}} \right), \left(\tfrac{2}{\sqrt{15}}, -\tfrac{1}{\sqrt{15}}, -\tfrac{1}{\sqrt{15}}, \tfrac{1}{\sqrt{15}} \right), \left(\tfrac{1}{\sqrt{3}}, 0, \tfrac{1}{\sqrt{3}}, -\tfrac{1}{\sqrt{3}} \right) \right\}$$

is an orthonormal basis for R^4 and find the coordinates of $(1,2,3,4)$ relative to the ordered basis S.

2. Find an orthonormal basis for the subspace of R^5 spanned by

$$\{(1,0,2,3,1),\ (1,0,2,1,3),\ (1,2,3,1,0)\}.$$

3. Find an orthonormal basis for the null space of $\begin{bmatrix} 2 & 3 & 1 & 4 & 1 & 1 \\ 2 & 4 & 1 & 2 & 1 & 2 \\ 2 & 3 & 1 & 1 & 2 & 2 \end{bmatrix}$.

4. Let $<\mathbf{u},\mathbf{v}>$ be the inner product on R^4 generated by $\begin{bmatrix} 1 & 2 & 1 & 3 \\ 2 & 1 & 3 & 2 \\ 1 & 1 & 1 & 2 \\ 3 & 2 & 1 & 3 \end{bmatrix}$. (See Chapter 11.)

   Find an orthonormal basis for R^4 using this inner product by applying the Gram-Schmidt process to the standard basis for R^4.

5. This is a variation of Solved Problem 2, where $<f,g>= \int_{-1}^{1} f(x)g(x)\,dx$.

   (a) Show that $\left\{ \frac{1}{\sqrt{2}},\ \cos(\pi x),\ \cos(2\pi x),\ \cos(3\pi x) \right\}$ is an orthonormal set.

   (b) Find the projection of x^2 onto the subspace spanned by these four functions.

   (c) On the same axes, graph x^2 and the projection of x^2 onto the subspace spanned by the new set S.

   (d) Find the projection of x onto the subspace spanned by these four functions. By examining the integrals, explain how you can predict the projections of x, x^3, x^5, ... without computing anything.

6. Let $<f,g>= \int_{-1}^{1} f(x)g(x)\,dx$. Using this inner product, find an orthonormal basis for the subspace of $C(-1,1)$ spanned by $\sin^2(\pi x)$, $\sin^2(2\pi x)$, and $\sin^2(3\pi x)$.

Exploration and Discovery

1. Try to produce a *DERIVE* function that will execute the Gram-Schmidt process in a single step. (Look at the QR.MTH file in Appendix II.)

2. This is a continuation of Solved Problem 2 where we found the projection of $f(x) = x$ onto the subspace spanned by S.

   (a) Find the distance between $f(x) = x$ and its projection.

   (b) Include $\sin(4\pi x)$ in the set S and then find the projection of $f(x) = x$ onto the subspace spanned by this new set.

   (c) Find the distance between $f(x) = x$ and its new projection. Is the approximation better than before? Explain.

   (d) On the same axes, graph x and the projection of x onto the subspace spanned by the new set S.

   (e) Repeat all the steps above for $\sin(5\pi x)$.

   (f) What do you think will happen if you continue to include $\sin(6\pi x)$, $\sin(7\pi x)$, and so on in the set S?

LABORATORY EXERCISE 12.1

Executing The Gram-Schmidt Process

Name ______________________ **Due Date** ______________

1. Find an orthonormal basis for the subspace of R^5 spanned by

$$\{(0,3,3,1,1),\ (1,0,0,1,2),\ (1,2,5,7,8)\}$$

   Carry out the procedure as we did in Solved Problem 3. Show the steps involved clearly.

2. Express each vector in the spanning set above as a linear combination of the orthonormal basis vectors you found.

CHAPTER 13

CHANGE OF BASIS AND ORTHOGONAL MATRICES

LINEAR ALGEBRA CONCEPTS
• **Transition (or change of basis) matrix** • **Orthogonal matrices**

Introduction

If B is an ordered basis for a vector space V then a vector $\mathbf{w}$ in V has a coordinate vector relative to B, which we will denote by $[\mathbf{w}]_B$. If B and B' are two ordered bases for V then $\mathbf{w}$ has a coordinate vector relative to both, and one can be obtained from the other by multiplication by a matrix called the *transition matrix* or *change of basis matrix.*

If V is an inner product space and if B and B' are orthonormal ordered bases, then the transition matrix has a special structure and is called an *orthogonal* matrix.

Theorem: For an n by n matrix A the following are equivalent.

1. A is orthogonal.
2. $AA^t = I$.
3. The rows of A form an orthonormal basis for R^n.
4. The columns of A form an orthonormal basis for R^n.
5. $||A\mathbf{v}|| = ||\mathbf{v}||$ for all $\mathbf{v}$ in R^n.

We used the equivalence of 2 and 3 above in Solved Problem 1 of Chapter 12. Number 5 above says that A preserves distance.

Solved Problems

Solved Problem 1: In R^4, let $B = \{\mathbf{u_1}, \mathbf{u_2}, \mathbf{u_3}, \mathbf{u_4}\}$, where $\mathbf{u_1} = (4,4,1,3)$, $\mathbf{u_2} = (3,2,2,1)$, $\mathbf{u_3} = (1,5,1,6)$, and $\mathbf{u_4} = (4,2,6,7)$. Let $B' = \{\mathbf{v_1}, \mathbf{v_2}, \mathbf{v_3}, \mathbf{v_4}\}$, where $\mathbf{v_1} = (1,2,3,4)$, $\mathbf{v_2} = (3,1,5,5)$, $\mathbf{v_3} = (3,1,5,1)$ and $\mathbf{v_4} = (2,1,1,2)$. Assume that B and B' are ordered bases. Let $\mathbf{w}$ be the vector whose coordinate vector relative to B is $[\mathbf{w}]_B = (4,5,6,7)$.

```
2:   u2 := [3, 2, 2, 1]

3:   u3 := [1, 5, 1, 6]

4:   u4 := [4, 2, 6, 7]

5:   v1 := [1, 2, 3, 4]

6:   v2 := [3, 1, 5, 5]

7:   v3 := [3, 1, 5, 1]

8:   v4 := [2, 1, 1, 2]

9:   4 u1 + 5 u2 + 6 u3 + 7 u4

10:  [65, 70, 62, 102]

11:  u1 = a v1 + b v2 + c v3 + d v4

12:  [4 = a + 3 b + 3 c + 2 d, 4 = 2 a + b + c + d, 1 = 3 a + 5 b + 5 c + d, 3 =

                     3       1
13:  [a = 1, b = - ———, c = ———, d = 3]
                     2       2
```

Figure 13.1: **Finding a transition matrix**

1. Find **w**.
2. Find the transition matrix from B to B'.
3. Find the coordinate vector of **w** relative to B'.

Solution 1: First we **Author** and name the eight vectors as in the first eight expressions in Figure 13.1. Since $[\mathbf{w}]_B = (4, 5, 6, 7)$, $\mathbf{w} = 4\mathbf{u}_1 + 5\mathbf{u}_2 + 6\mathbf{u}_3 + 7\mathbf{u}_4$. Thus we **Author** and **Simplify** `4u1+5u2+6u3+7u4`. The vector **w** is displayed in expression 10 of Figure 13.1.

The transition matrix from B to B' is the matrix whose columns are $[\mathbf{u}_1]_{B'}$, $[\mathbf{u}_2]_{B'}$, $[\mathbf{u}_3]_{B'}$, $[\mathbf{u}_4]_{B'}$. Thus we need to find the B' coordinate vectors of $\mathbf{u}_1, \mathbf{u}_2, \mathbf{u}_3$ and $\mathbf{u}_4$.

To find $[\mathbf{u}_1]_{B'}$ we need to solve the equation $\mathbf{u}_1 = a\mathbf{v}_1 + b\mathbf{v}_2 + c\mathbf{v}_3 + d\mathbf{v}_4$ for the variables a, b, c, d. **Author** `[u1=av1 +bv2 +cv3+ dv4]` and **Simplify**; then **soLve** to get the solution in expression 13 of Figure 13.1. These values make up the first column of the transition matrix, which appears in expression 16 of Figure 13.2.

$$
16: \begin{bmatrix} 1 & \frac{4}{13} & \frac{32}{13} & \frac{4}{13} \\ -\frac{3}{2} & -\frac{10}{13} & -\frac{73}{52} & \frac{51}{52} \\ \frac{1}{2} & \frac{9}{13} & -\frac{11}{52} & -\frac{3}{52} \\ 3 & \frac{19}{13} & \frac{22}{13} & \frac{6}{13} \end{bmatrix} \cdot [4, 5, 6, 7]
$$

$$
17: \left[\frac{292}{13}, -\frac{593}{52}, \frac{197}{52}, \frac{425}{13}\right]
$$

$$
18: \frac{292}{13} v1 - \frac{593}{52} v2 + \frac{197}{52} v3 + \frac{425}{13} v4
$$

19: [65, 70, 62, 102]

Figure 13.2: **Finding the coordinates of w relative to B'**

Repeat this procedure for each of the others: $[\mathbf{u}_2]_{B'}$, $[\mathbf{u}_3]_{B'}$ and $[\mathbf{u}_4]_{B'}$ to complete the transition matrix in expression 16 of Figure 13.2.

The B' coordinates of $\mathbf{w}$ can be found from $[\mathbf{w}]_{B'} = M[\mathbf{w}]_B$ where M is the transition matrix. Assuming that the transition matrix has been named M (which is not shown in the figures) we **Author** `M.[4,5,6,7]`, which is seen as expression 16. **Simplify** it, and we get the B' coordinate vector of $\mathbf{w}$ in expression 17 of Figure 13.2. Observe that in expressions 18 and 19, we have checked the answer and it agrees with that for $\mathbf{w}$, as in expression 10 of Figure 13.1.

Solution 2: The method we presented in solution 1 is intended to show a step-by-step analysis, but it is tedious because we have to enter and solve four systems of equations and then put the results in a matrix. There is a more efficient way: One can show that if C is the matrix whose columns are the vectors $\mathbf{u}_1, \mathbf{u}_2, \mathbf{u}_3, \mathbf{u}_4$ and D is the matrix whose columns are the vectors $\mathbf{v}_1, \mathbf{v}_2, \mathbf{v}_3, \mathbf{v}_4$ then the four systems of equations we had to solve are equivalent to the single matrix equation $C = DA$. Its solution is $A = D^{-1}C$. (Explain

$$3: \quad \begin{bmatrix} 1 & 3 & 3 & 2 \\ 2 & 1 & 1 & 1 \\ 3 & 5 & 5 & 1 \\ 4 & 5 & 1 & 2 \end{bmatrix}^{-1} \cdot \begin{bmatrix} 4 & 3 & 1 & 4 \\ 4 & 2 & 5 & 2 \\ 1 & 2 & 1 & 6 \\ 3 & 1 & 6 & 7 \end{bmatrix}$$

$$4: \quad \begin{bmatrix} 1 & \frac{4}{13} & \frac{32}{13} & \frac{4}{13} \\ -\frac{3}{2} & -\frac{10}{13} & -\frac{73}{52} & \frac{51}{52} \\ \frac{1}{2} & \frac{9}{13} & -\frac{11}{52} & -\frac{3}{52} \\ 3 & \frac{19}{13} & \frac{22}{13} & \frac{6}{13} \end{bmatrix}$$

Figure 13.3: **Alternative solution to Solved Problem 1**

why D has an inverse.)

(You may use **Transfer Clear** to refresh your screen before continuing.) It is relatively efficient to use **Declare Matrix** to enter $C = \begin{bmatrix} 4 & 3 & 1 & 4 \\ 4 & 2 & 5 & 2 \\ 1 & 2 & 1 & 6 \\ 3 & 1 & 6 & 7 \end{bmatrix}$ as expression 1 and $D = \begin{bmatrix} 1 & 3 & 3 & 2 \\ 2 & 1 & 1 & 1 \\ 3 & 5 & 5 & 1 \\ 4 & 5 & 1 & 2 \end{bmatrix}$ as expression 2 (not shown). Next **Author** **`#2^(-1).#1`** as in expression 3 of Figure 13.3 and **Simplify** to get A as expression 4. Observe that this agrees with the earlier calculation of A that appears in expression 16 of Figure 13.2.

<u>**Solved Problem 2**</u>: Show that $A = \begin{bmatrix} \frac{}{2} & \frac{1}{2} & \frac{1}{2} & \frac{1}{2} \\ 0 & -\frac{1}{\sqrt{2}} & \frac{1}{\sqrt{2}} & 0 \\ -\frac{5}{6} & \frac{1}{6} & \frac{1}{6} & \frac{1}{2} \\ \frac{\sqrt{2}}{6} & -\frac{\sqrt{2}}{3} & -\frac{\sqrt{2}}{3} & \frac{1}{\sqrt{2}} \end{bmatrix}$ is orthogonal (a) by showing that $||A\mathbf{v}|| = ||\mathbf{v}||$ for all $\mathbf{v}$ and (b) by another method.

<u>Solution</u>: If you find this matrix tedious to enter using **Declare Matrix**, you may wish to look back to Solved Problem 1 in Chapter 12 where we suggested in detail how to handle such an example. We will assume that the reader has successfully entered the matrix and named it A.

We **Author** `|A.[x,y,z,w]|` as expression 6 of Figure 13.4 and **Simplify** it to obtain expression 7, which we recognize as $||(x, y, z, w)||$. This is all we have to do to verify part (a), but to show the complexity of what we just did, we will **Author** and **Simplify** `A.[x,y,z,w]` as well. The resulting expression 9 of Figure 13.4 is quite complicated and runs off the screen. It is not at all obvious that the length of this huge vector is the same as that of (x, y, z, w).

Now for part (b): The simplest way to verify that A is orthogonal is to show that $AA^t = I$ (number 2 in the Theorem in the Introduction). Therefore, we **Author** `A.A'` as expression 10 of Figure 13.4. When we **Simplify** it we see the identity matrix and conclude that A is orthogonal.

6: $|a \cdot [x, y, z, w]|$

7: $\sqrt{x^2 + y^2 + z^2 + w^2}$

8: $a \cdot [x, y, z, w]$

9: $\left[\frac{x}{2} + \frac{y}{2} + \frac{z}{2} + \frac{w}{2}, \frac{\sqrt{2}\, z}{2} - \frac{\sqrt{2}\, y}{2}, -\frac{5\, x}{6} + \frac{y}{6} + \frac{z}{6} + \frac{w}{2}, \frac{\sqrt{2}\, x}{6}\right.$

10: $a \cdot a'$

11: $\begin{bmatrix} 1 & 0 & 0 & 0 \\ 0 & 1 & 0 & 0 \\ 0 & 0 & 1 & 0 \\ 0 & 0 & 0 & 1 \end{bmatrix}$

Figure 13.4: **An orthogonal matrix preserves length**

Exercises

1. Assume that $B = \{(2,3,4,1),(4,4,5,2),(1,2,3,3),(1,1,1,1)\}$ and $B' = \{(2,7,5,3),(1,4,2,2),(5,1,3,2),(7,7,4,5)\}$ are ordered bases for R^4. Suppose that the B coordinate vector of $\mathbf{w}$ is $[\mathbf{w}]_B = (4,1,3,1)$.

   (a) Find $\mathbf{w}$.

   (b) Find the transition matrix from B to B'.

   (c) Find the B' coordinate vector $[\mathbf{w}]_{B'}$ of $\mathbf{w}$.

   (d) Find the transition matrix from B' to B.

   (e) Multiply the matrices in (b) and (d). What do you conclude about them?

2. $B = \{x^3 - 2x^2 + 1, x^3 + x - 3, x^2 + 3x - 8, x^3 + 4x^2 - x - 1\}$ and $B' = \{x^3 - 3x^2 + 2x, x^3 + x^2 + x + 1, x^3 + 7x - 3, x^3 + 4x^2 - 3x + 4\}$ are ordered bases for P_3. $p(x)$ is the cubic polynomial whose B coordinate vector is $[p(x)]_B = (2,4,1,5)$.

   (a) Find $p(x)$.

   (b) Find the transition matrix from B to B'.

   (c) Find the B' coordinate vector, $[p(x)]_{B'}$, of $p(x)$.

   (d) Find the transition matrix from B' to B.

   (e) Multiply the matrices in (b) and (d). What do you conclude about them?

3. $B = \left\{ \begin{bmatrix} 2 & 3 \\ 1 & 2 \end{bmatrix}, \begin{bmatrix} 3 & 6 \\ 1 & 1 \end{bmatrix}, \begin{bmatrix} 4 & 2 \\ 8 & 3 \end{bmatrix}, \begin{bmatrix} 1 & 3 \\ 4 & 2 \end{bmatrix} \right\}$ and $B' = \left\{ \begin{bmatrix} 3 & 7 \\ 2 & 5 \end{bmatrix}, \begin{bmatrix} 4 & 5 \\ 2 & 2 \end{bmatrix}, \begin{bmatrix} 5 & 2 \\ 1 & 9 \end{bmatrix}, \begin{bmatrix} 3 & 3 \\ 2 & 6 \end{bmatrix}, \begin{bmatrix} 1 & 1 \\ 5 & 8 \end{bmatrix} \right\}$ are ordered bases for $M_{2,2}$. $\mathbf{w}$ is the matrix whose B coordinate vector is $[\mathbf{w}]_B = (4,1,3,3)$.

   (a) Find $\mathbf{w}$.

   (b) Find the transition matrix from B to B'.

(c) Find the B' coordinate vector, $[\mathbf{w}]_{B'}$, of $\mathbf{w}$.

(d) Find the transition matrix from B' to B.

(e) Multiply the matrices in (b) and (d). What do you conclude about them?

4. $B = \{(1,3,2,4),(4,2,3,3),(5,5,3,4),(1,8,1,3)\}$ is an ordered basis for R^4. Suppose B' is another ordered basis for R^4 and that the transition matrix from B to B' is $\begin{bmatrix} 1 & 3 & 2 & 5 \\ 6 & 7 & 1 & 1 \\ 4 & 2 & 3 & 3 \\ 1 & 8 & 2 & 2 \end{bmatrix}$.

Find B'.

Exploration and Discovery

1. Use the examples of 4 by 4 orthogonal matrices below to test the following propositions. Provide counterexamples for those that are false, and prove the ones that are true.

$$\begin{bmatrix} \frac{1}{2} & -\frac{1}{2} & \frac{1}{2} & \frac{1}{2} \\ -\frac{\sqrt{2}}{6} & \frac{1}{\sqrt{2}} & \frac{\sqrt{2}}{3} & \frac{\sqrt{2}}{3} \\ \frac{5\sqrt{2}}{9} & \frac{\sqrt{2}}{3} & -\frac{5\sqrt{2}}{18} & \frac{\sqrt{2}}{18} \\ -\frac{5}{18} & -\frac{1}{6} & -\frac{11}{18} & \frac{13}{18} \end{bmatrix} \begin{bmatrix} \frac{1}{\sqrt{3}} & 0 & \frac{1}{\sqrt{3}} & -\frac{1}{\sqrt{3}} \\ -\frac{1}{\sqrt{3}} & \frac{1}{\sqrt{3}} & \frac{1}{\sqrt{3}} & 0 \\ \frac{2}{\sqrt{15}} & \frac{1}{\sqrt{15}} & \frac{1}{\sqrt{15}} & \frac{1}{\sqrt{15}} \\ -\frac{1}{\sqrt{15}} & -\frac{3}{\sqrt{15}} & \frac{2}{\sqrt{15}} & \frac{1}{\sqrt{15}} \end{bmatrix} \begin{bmatrix} -\frac{1}{2} & \frac{1}{2} & \frac{1}{2} & -\frac{1}{2} \\ 0 & \frac{1}{\sqrt{6}} & \frac{1}{\sqrt{6}} & \frac{2}{\sqrt{6}} \\ \frac{6}{\sqrt{66}} & -\frac{1}{\sqrt{66}} & \frac{5}{\sqrt{66}} & -\frac{2}{\sqrt{66}} \\ -\frac{3\sqrt{11}}{22} & -\frac{5\sqrt{11}}{22} & \frac{3\sqrt{11}}{22} & \frac{\sqrt{11}}{22} \end{bmatrix}$$

   (a) If A is orthogonal, then A^{-1} is orthogonal.

   (b) If A and B are orthogonal, then AB is orthogonal.

   (c) If A and B are orthogonal, then $A + B$ is orthogonal.

   (d) If A is orthogonal, then $\text{DET}(A) = 1$.

2. Discuss how to generate examples of 5 by 5 orthogonal matrices (other than diagonal ones). Make up a few and check them. (Hint: You might make examples that have a lot of zeros. The Gram-Schmidt process may help you to find more interesting examples.)

LABORATORY EXERCISE 13.1

Transition Matrices

Name ____________________ Due Date ____________________

Suppose that $B = \{(1,2,3,1), (2,3,1,2), (1,1,2,3), (1,2,1,2)\}$ and $B' = \{(2,1,1,3), (1,1,2,2), (2,1,3,3), (1,1,1,1)\}$ are bases for R^4, and that the coordinate vector of $\mathbf{w}$ relative to B is $[\mathbf{w}]_B = (1,2,3,4)$.

1. Find $\mathbf{w}$.

2. Find the transition matrix from B to B'.

3. Find the B' coordinate vector $[\mathbf{w}]_{B'}$ of $\mathbf{w}$.

4. Find the transition matrix from B' to B.

5. Find the inverse of the matrix in Part 2. Compare the answer to that found in Part 4.

CHAPTER 14

EIGENVALUES AND EIGENVECTORS

LINEAR ALGEBRA CONCEPTS

- **Characteristic polynomial**
- **Eigenvalue**
- **Eigenvector**
- **Eigenspace**

Introduction

DERIVE's syntax for the characteristic polynomial of a matrix A is CHARPOLY(A). *DERIVE* can find the exact eigenvalues and eigenvectors of many matrices, but often both eigenvalues and eigenvectors must be approximated. The VECTOR.MTH file that comes with the *DERIVE* program contains the tools necessary to do this.

In many texts the Greek letter λ (lambda) is used for eigenvalues. *DERIVE* supports several Greek characters, but not this one. Therefore another letter must be used. Since *DERIVE*'s default variable for characteristic polynomials is w, that is what we will use.

Solved Problems

Solved Problem 1: Let $A = \begin{bmatrix} 45 & -21 & -63 & -12 & 21 \\ 28 & -12 & -42 & -7 & 14 \\ 31 & -15 & -43 & -9 & 15 \\ 28 & -14 & -42 & -5 & 14 \\ 51 & -25 & -75 & -14 & 27 \end{bmatrix}$.

1. Find the characteristic polynomial of A.
2. Find the Eigenvalues of A.
3. Find a basis for each eigenspace of A.

Solution: Enter the matrix using **Declare Matrix**, and name it A. It will be convenient for part 3 to have a short name for the 5 by 5 identity matrix. **Author** `I:=identity_matrix(5)`.

2: $a := \begin{bmatrix} 45 & -21 & -63 & -12 & 21 \\ 28 & -12 & -42 & -7 & 14 \\ 31 & -15 & -43 & -9 & 15 \\ 28 & -14 & -42 & -5 & 14 \\ 51 & -25 & -75 & -14 & 27 \end{bmatrix}$

3: i := IDENTITY_MATRIX (5)

4: CHARPOLY (a)

5: $-w^5 + 12\,w^4 - 57\,w^3 + 134\,w^2 - 156\,w + 72$

6: $w = 2$

7: $w = 3$

Figure 14.1: **Eigenvalues of a matrix**

To find the characteristic polynomial, we may **Author** and **Simplify** either `det(wI-A)` or `charpoly(A)`. The characteristic polynomial appears in expression 5 of Figure 14.1.

To find the eigenvalues, we **soLve** the characteristic polynomial. The eigenvalues 2 and 3 appear in expressions 6 and 7 of Figure 14.1.

The eigenspace corresponding to 2 is the null space of $2I - A$. Thus we **Author** and **Simplify** `row_reduce(2I-A)`. The reduced matrix appears in expression 9 of Figure 14.2. From this matrix we read a basis for the null space (and hence for the eigenspace belonging to the eigenvalue 2) as seen below.

$$\begin{bmatrix} 0 \\ -3 \\ 1 \\ 0 \\ 0 \end{bmatrix}, \begin{bmatrix} \frac{3}{2} \\ \frac{5}{2} \\ 0 \\ 1 \\ 0 \end{bmatrix}, \begin{bmatrix} 0 \\ 1 \\ 0 \\ 0 \\ 1 \end{bmatrix}$$

In the same way, **Author** and **Simplify** `row_reduce(3I-A)` to get the basis for the null space displayed below.

$$\begin{bmatrix} -\frac{2}{7} \\ 1 \\ -\frac{5}{7} \\ 1 \\ 0 \end{bmatrix}, \begin{bmatrix} 1 \\ 0 \\ 1 \\ 0 \\ 1 \end{bmatrix}$$

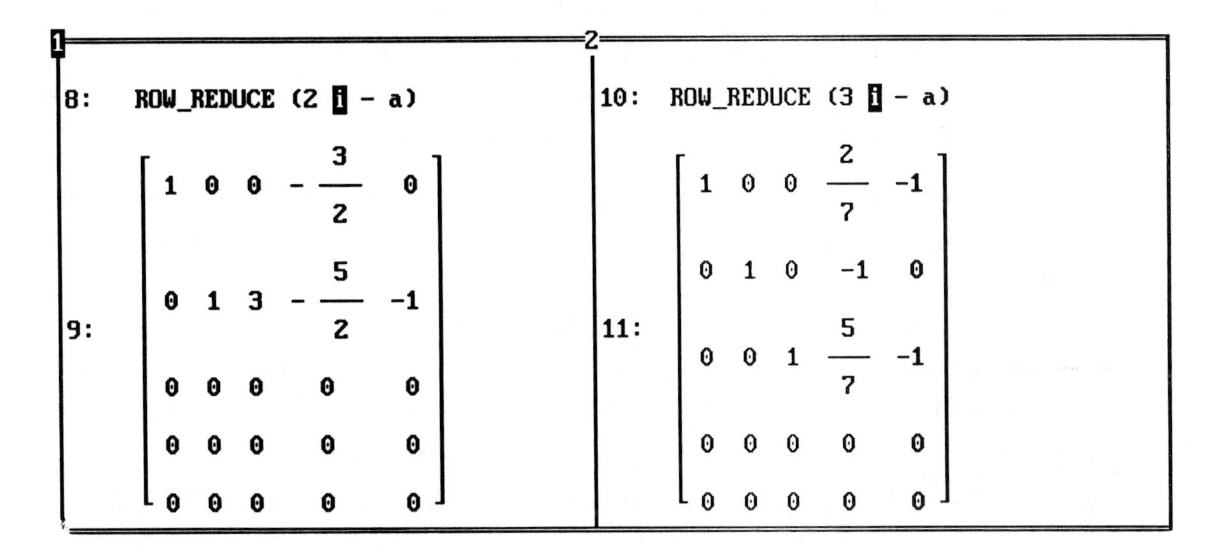

Figure 14.2: **Eigenvectors of a matrix**

<u>Solved Problem 2</u>: Find the eigenvalues and associated eigenvectors of $A = \begin{bmatrix} 3 & 2 & 5 \\ 3 & 4 & 1 \\ 6 & 7 & 3 \end{bmatrix}$.

<u>Solution</u>: The eigenvalues are found exactly as in Solved Problem 1. One of the eigenvalues is the extremely complicated expression 5 of Figure 14.3. When we seek its corresponding eigenvectors, serious problems occur as we will see next when we try to row reduce $wI - A$.

Highlight the eigenvalue in expression 5. *Be sure to use the right arrow key to highlight the right side of the equal sign, excluding the "$w =$".* Now **Author** and **Simplify** `row_reduce(`F4`identity_matrix(3)-A)`. *DERIVE* produces the 3 by 3 identity matrix

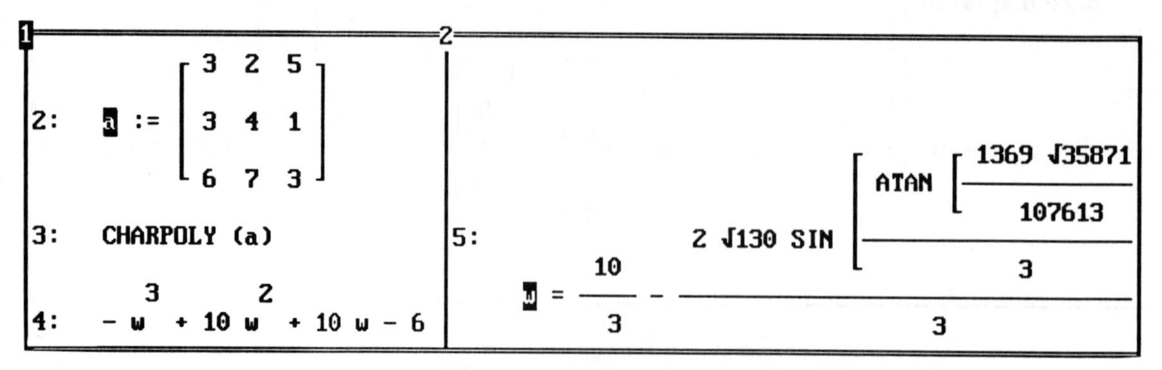

Figure 14.3: **A matrix with complicated eigenvalues**

in expression 9 of Figure 14.4. This is incorrect, of course, because an eigenvalue is a number w that makes the matrix $wI - A$ singular; it cannot reduce to the identity. What's going on?

When *DERIVE* tries to row-reduce the matrix $wI - A$, where w is as in expression 5, it encounters entries that are actually equivalent to zero but are so complicated that *DERIVE* can't recognize them. At some point division by zero occurs, and an incorrect reduced matrix is returned. Thus it is necessary to settle for approximate eigenvectors.

Load the VECTOR.MTH file into memory using **Transfer Load Utility** VECTOR. If w is an approximate eigenvalue (*w must not be an exact eigenvalue*) for a matrix A, then **Author** and **approX** `approx_eigenvector(A,w)` to return an approximate eigenvector associated with w. Highlight expression 5 of Figure 14.3 and use **approX** to get an approximate value in expression 10 of Figure 14.4. Next **Author** and **approX**

```
approx_eigenvector(A,0.426144)
```

The approximate eigenvector is expression 14 in Figure 14.4.

What does "approximate" mean? Just how close is our result to a true eigenvector? *DERIVE* gives no way to tell, but the following idea gives us some "evidence" of nearness:

If w is a true eigenvalue and $\mathbf{v}$ a corresponding true eigenvector, then $A\mathbf{v} = w\mathbf{v}$. Therefore, if the eigenvalue and corresponding eigenvector are good approximations, the two sides of the equation should be very close to each other.

Let's test this: **Author** and **approX** `[A.#14, 0.426144(#14)]` as in expression 14 in Figure 14.4. A times the approximate eigenvector appears in the first row of expression 16 and

```
                                                  ┌        ┌ 1369 √35871 ┐ ┐ ┐
                                                  │   ATAN │─────────────│ │ │
                                                  │        └   107613    ┘ │ │
8:                                  2 √130 SIN    │────────────────────────│ │
                      ┌┌   10                     └            3           ┘ │
         ROW_REDUCE   ││  ──── - ─────────────────────────────────────────────│ IDENTITY_MATRIX
                      └└   3                            3                    ┘

      ┌ 1  0  0 ┐
9:    │ 0  1  0 │
      └ 0  0  1 ┘

10:  w = 0.426144

13:  APPROX_EIGENVECTOR (a, 0.426144)

14:  [0.777612, -0.608894, -0.156735]

15:  [a · [0.777612, -0.608894, -0.156735], 0.426144 [0.777612, -0.608894, -0.15

      ┌ 0.331372  -0.259475  -0.0667910 ┐
16:   │                                 │
      └ 0.331374  -0.259476  -0.0667916 ┘
```

Figure 14.4: **Approximating eigenvectors**

0.42611 times the approximate eigenvector appears in the second row. These two vectors agree through five decimal places, so we are led to suspect that (0.777612, −0.608894, −0.156735) is very near to some true eigenvector belonging to the exact eigenvalue in expression 5 of Figure 14.3. But this is just evidence. We cannot conclude anything quantitatively about *how* near. It depends on A and *DERIVE* gives no way to estimate it.

We leave it to the reader to approximate and check the eigenvectors for the remaining two eigenvalues in a similar way.

> **Summary**: If you find an eigenvalue w for which *DERIVE* row-reduces $wI - A$ to the identity matrix, then you must **approXimate** w and its eigenvectors. To do so, use **Transfer Load Utility** and type the file name VECTOR.MTH. After it is loaded, **Author** and **approX approx_eigenvector**(A, w).

Solved Problem 3: Approximate the eigenvalues and eigenvectors of $\begin{bmatrix} 1 & 1 & 2 & 1 & 1 \\ 1 & 2 & 3 & 2 & 1 \\ 2 & 3 & 1 & 2 & 1 \\ 1 & 2 & 2 & 3 & 1 \\ 1 & 1 & 1 & 1 & 7 \end{bmatrix}$.

Solution: Enter the 5 by 5 matrix and name it A. Next **Author** and **Simplify** `charpoly(A)`. The characteristic polynomial appears in expression 4 of Figure 14.5. If we ask *DERIVE* to **soLve** this, it will fail. Thus we must be satisfied with an approximation of the zeros of the characteristic polynomial. With expression 4 highlighted, use **Plot Overlay Plot** to make rough estimates of the zeros. (See Section 13 of Appendix I for details on plotting.) The graph in Figure 14.6 uses the **Scale** `x:3, y:250`. (Use the Tab key to set the y scale.) From the graph we see that there are 5 zeros lying in the respective intervals $(-3,0),(0,1),(1,3),(3,6),(9,12)$. Use **Algebra** to return to the calculations and set *DERIVE* to its approximate mode using **Options Precision Approximate** Enter.

2: $a := \begin{bmatrix} 1 & 1 & 2 & 1 & 1 \\ 1 & 2 & 3 & 2 & 1 \\ 2 & 3 & 1 & 2 & 1 \\ 1 & 2 & 2 & 3 & 1 \\ 1 & 1 & 1 & 1 & 7 \end{bmatrix}$

3: CHARPOLY (a)

4: $-w^5 + 14w^4 - 39w^3 - 59w^2 + 121w - 31$

5: w = -1.89843

6: w = 0.312619

7: w = 1.0712

8: w = 5.28071

9: w = 9.23388

Figure 14.5: **Approximating eigenvalues of a matrix**

Finally, we ask *DERIVE* to **soLve** on each of the intervals listed above. See Section 11 in Appendix I for details on solving equations. The approximate eigenvalues are expressions 5 through 9 of Figure 14.5. The eigenvectors should be approximated using the APPROX_EIGENVECTOR function and checked for accuracy as in Solved Problem 2.

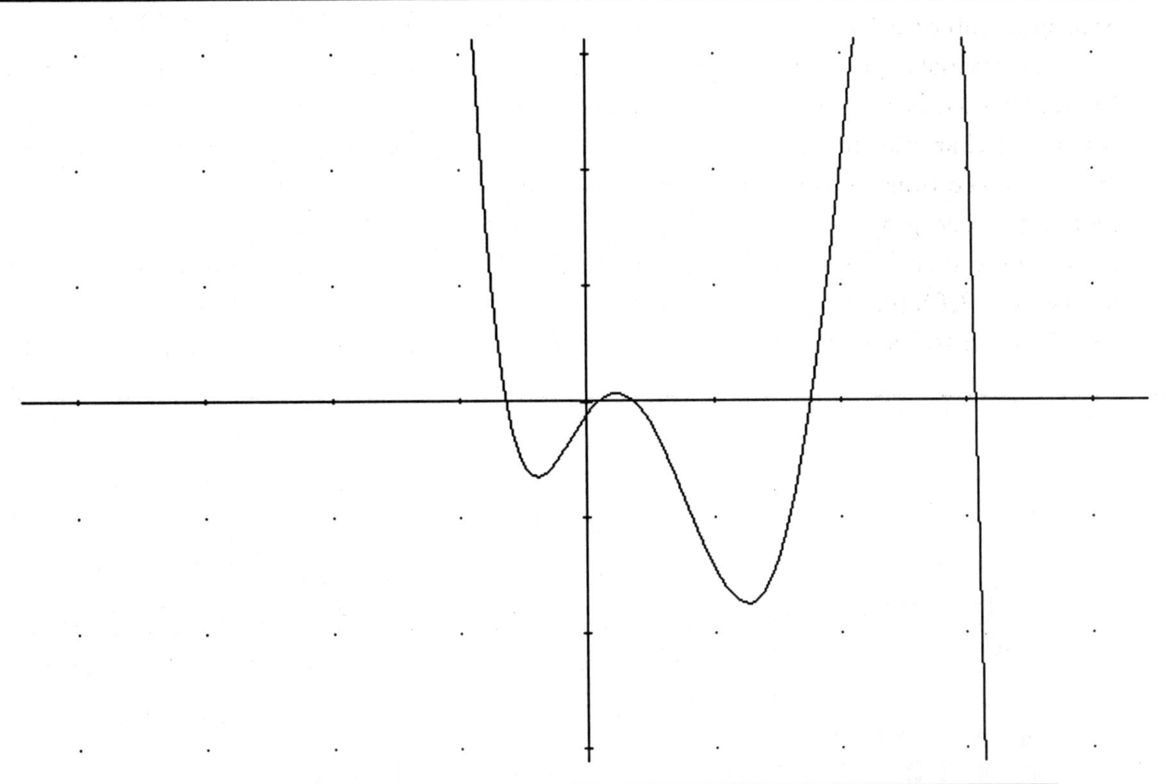

Figure 14.6: **The graph of a characteristic polynomial. Scale 3:250**

Exercises

1. For each of the following matrices, find the characteristic polynomial, the exact eigenvalues, and a basis for each eigenspace, exactly.

$$A = \begin{bmatrix} 3 & 4 & -4 & -4 \\ 4 & 3 & -4 & -4 \\ 0 & 4 & -1 & -4 \\ 4 & 0 & -4 & -1 \end{bmatrix} \qquad B = \begin{bmatrix} 0 & 4 & 4 & -4 \\ -6 & 10 & 6 & -6 \\ -2 & 2 & 6 & -2 \\ -6 & 6 & 6 & -2 \end{bmatrix}$$

2. For each of the following matrices:

   (a) Find the characteristic polynomial.

   (b) Approximate the eigenvalues and report their exact values where it is practical to do so, that is, if they can be found exactly and are not immensely complicated.

   (c) Find a basis for each eigenspace. Where it is practical to do so report the eigenvectors exactly. Otherwise approximate them.

$$A = \begin{bmatrix} -2 & 4 & 4 & 3 \\ -2 & 9 & 3 & 0 \\ -1 & 4 & 2 & 1 \\ 1 & 4 & 2 & 1 \end{bmatrix} \quad B = \begin{bmatrix} -5 & 11 & 0 & -1 \\ -4 & 12 & 1 & 0 \\ -4 & 9 & 0 & -2 \\ -2 & 8 & 0 & 1 \end{bmatrix} \quad C = \begin{bmatrix} -2 & 5 & 4 & 0 \\ -3 & 9 & 3 & -1 \\ -1 & 3 & 4 & 0 \\ -2 & 7 & 1 & -1 \end{bmatrix}$$

$$D = \begin{bmatrix} -2 & -4 & 8 & -2 & 8 \\ -16 & 2 & 18 & 6 & 0 \\ -2 & -4 & 8 & -3 & 8 \\ 8 & -6 & -4 & -6 & 12 \\ -2 & -4 & 6 & -3 & 10 \end{bmatrix} \quad E = \begin{bmatrix} -3 & 8 & 0 & 8 & -1 \\ 0 & 12 & -4 & 10 & -3 \\ -3 & 8 & 1 & 9 & 1 \\ -2 & 1 & 3 & 3 & 2 \\ -4 & 7 & 3 & 6 & -1 \end{bmatrix} \quad F = \begin{bmatrix} -9 & 2 & 7 & 2 & 5 \\ -11 & 0 & 9 & 0 & 7 \\ -21 & 2 & 17 & 1 & 8 \\ -15 & 2 & 12 & 1 & 3 \\ -10 & 1 & 10 & -1 & 4 \end{bmatrix}$$

3. For each of the following matrices, find the characteristic polynomial. Approximate the eigenvalues and approximate the eigenvectors forming a basis for the eigenspace corresponding to each eigenvalue.

$$A = \begin{bmatrix} 12 & 0 & -7 & -7 & 7 \\ 10 & 1 & -5 & -6 & 6 \\ 7 & 1 & -4 & -4 & 4 \\ 12 & -1 & -7 & -7 & 6 \\ 8 & 1 & -4 & -4 & 3 \end{bmatrix} \quad B = \begin{bmatrix} 4 & 3 & -1 & -4 & 5 \\ 0 & 5 & 1 & 0 & 3 \\ 7 & 1 & -3 & -5 & 5 \\ 2 & 3 & 0 & -3 & 3 \\ 3 & 3 & 0 & -2 & 2 \end{bmatrix} \quad C = \begin{bmatrix} 21 & -1 & -10 & -13 & 12 \\ 18 & -3 & -5 & -11 & 11 \\ 13 & 2 & -6 & -9 & 8 \\ 17 & -1 & -9 & -10 & 9 \\ 19 & -1 & -9 & -12 & 10 \end{bmatrix}$$

4. Assume that A is a 4 by 4 matrix with eigenvalues 2, 3, and 4. Bases for the corresponding eigenspaces are given in the following table. Find A.

Eigenvalue	4	3	2
Basis for Eigenspace	(3,1,2,1)	(4,4,2,3)	(2,1,1,1), (3,2,2,1)

Hint: Let A be a matrix with variable entries, and look at the following system of equations.

$$A\begin{bmatrix}2\\1\\1\\1\end{bmatrix}=2\begin{bmatrix}2\\1\\1\\1\end{bmatrix},\quad A\begin{bmatrix}3\\2\\2\\1\end{bmatrix}=2\begin{bmatrix}3\\2\\2\\1\end{bmatrix},\quad A\begin{bmatrix}4\\4\\2\\3\end{bmatrix}=3\begin{bmatrix}4\\4\\2\\3\end{bmatrix},\quad A\begin{bmatrix}3\\1\\2\\1\end{bmatrix}=4\begin{bmatrix}3\\1\\2\\1\end{bmatrix}.$$

Exploration and Discovery

1. We will explore the eigenvalues of functions of the matrix A of Exercise 1.

   (a) Find the eigenvalues of the matrix A of Exercise 1.

   (b) Find the eigenvalues of A^2 and A^3.

   (c) Find the eigenvalues of $A^2 + A^3$.

   (d) Find the eigenvalues of $p(A)$, where $p(x) = 3x^3 - 2x^2 + 1$.

   (e) Explain how your answers to (b) through (d) relate to your answer in (a). Formulate a general conjecture and try to prove it.

2. **(The Cayley-Hamilton Theorem)** For each of the three matrices in Exercise 1 calculate the characteristic polynomial $p(x)$ and evaluate $p(x)$ at the matrix. (See Chapter 3 to recall what this means.) Do Exploration and Discovery Problem 2 in Chapter 5.

3. Calculate the characteristic polynomials of A, B, and C in Exercise 2, and of their transposes. Discuss your observations and conclusions.

4. If A denotes the matrix from Exercise 2 and $\mathbf{v} = [1, 1, 1, 1]$, **approX** $A^5\mathbf{v}$, $A^{10}\mathbf{v}$, $A^{50}\mathbf{v}$, and $A^{100}\mathbf{v}$. Compare the vectors you obtain with the eigenvectors of A. Repeat the experiment using different (nonzero) vectors $\mathbf{v}$ and matrices B and C from Exercise 2. Report your findings.

5. The *trace* of a square matrix is defined to be the sum of its diagonal entries. Thus the

   trace of $\begin{bmatrix} -14 & -24 & -19 \\ 1 & 5 & 1 \\ 14 & 18 & 19 \end{bmatrix}$ is 10.

   (a) For the matrix above compute the trace and the characteristic polynomial. Make up a 4 by 4, a 5 by 5 and a new 3 by 3 matrix and do the same.

   (b) Compare the characteristic polynomials and the traces above. What do you notice?

   (c) Enter a 2 by 2 and a 3 by 3 matrix all of whose entries are variables. Compare their characteristic polynomials and traces. What do you notice?

(d) Use the **soLve** command to find the roots of the two characteristic polynomials of the 3 by 3 examples you used in (a) and (b). Find the sum of these roots. What do you notice?

(e) Prove that the coefficient of x^{n-1} in a polynomial $x^n + a_{n-1}x^{n-1} + \ldots + a_1 x + a_0$ is the sum of the roots of the polynomial. (Hint: Every such polynomial factors as $(x - r_1)(x - r_2)\ldots(x - r_n)$. The r_i's may be complex or may be repeated. Use *DERIVE* to **Expand** such products.)

(f) Summarize your observations and conclusions from all the above. What do your observations say about the relationship between the eigenvalues and the trace of a matrix?

(g) Show that if A is similar to B then they have the same trace. (Hint: Show similar matrices have the same determinant. Show $A - wI$ and $B - wI$ are similar and hence A and B have the same characteristic polynomial.)

LABORATORY EXERCISE 14.1

Eigenvalues and Eigenvectors

Name ______________________________ **Due Date** ______________

Let $A = \begin{bmatrix} 1 & -2 & 4 & 2 \\ -2 & 1 & 4 & 2 \\ 0 & -2 & 5 & 2 \\ -2 & -2 & 4 & 5 \end{bmatrix}$.

1. Find the characteristic polynomial of A.

2. Find the exact eigenvalues of A.

3. Find a basis for each eigenspace of A.

CHAPTER 15

DIAGONALIZATION AND ORTHOGONAL DIAGONALIZATION

LINEAR ALGEBRA CONCEPTS

- **Similarity**
- **Diagonalization**
- **Symmetric matrix**
- **Orthogonal diagonalization**

Introduction

Two matrices A and B are called *similar* if there is some invertible matrix P such that $B = P^{-1}AP$.

An n by n matrix is called *diagonalizable* if it is similar to a diagonal matrix. This is the case if it has a set of eigenvectors that forms a basis for R^n. In many instances, a solution to a problem for diagonal matrices can be carried over to diagonalizable matrices.

Solved Problems

Solved Problem 1: Determine if the following matrices are diagonalizable. If they are, exhibit the diagonalization.

$$A = \begin{bmatrix} 6 & 4 & -6 & 0 & 2 \\ 2 & 5 & -4 & 0 & 2 \\ 4 & 5 & -5 & 0 & 3 \\ 2 & 3 & -4 & 2 & 2 \\ 2 & 3 & -4 & -2 & 6 \end{bmatrix} \quad B = \begin{bmatrix} -6 & -5 & 12 & -9 & 2 \\ -4 & 1 & 5 & -2 & -1 \\ -8 & -3 & 13 & -7 & 0 \\ -4 & -1 & 5 & 0 & -1 \\ -4 & -1 & 5 & -1 & 0 \end{bmatrix}$$

Solution for A: Enter the 5 by 5 matrix using **Declare Matrix** and name it A. **Author** `I:=identity_matrix(5)`. To find the eigenvalues **Author** and **Simplify** `charpoly(A)`. Rather than **soLve** the characteristic polynomial, we **Factor** it. This shows not only the eigenvalues but their orders as well. From expression 6 of Figure 15.1 we see that

```
                 ┌ 6  4  -6   0  2 ┐
                 │                  │
                 │ 2  5  -4   0  2  │
                 │                  │
2:    a :=       │ 4  5  -5   0  3  │
                 │                  │
                 │ 2  3  -4   2  2  │
                 │                  │
                 └ 2  3  -4  -2  6 ┘

3:    i := IDENTITY_MATRIX (5)

4:    CHARPOLY (a)

          5        4        3         2
5:    - w  + 14 w  - 77 w  + 208 w  - 276 w + 144

                         2        2
6:    (4 - w) (w - 3)  (w - 2)
```

Figure 15.1: **Factored form of the characteristic polynomial of A**

4 is an eigenvalue of order 1 and that 3 and 2 are eigenvalues of order 2. In order to find the eigenvectors **Author** and **Simplify** `row_reduce(4I-A)`, `row_reduce(3I-A)`, and `row_reduce(2I-A)` as separate expressions. From expressions 8, 10 and 12 of Figure 15.2 we deduce the following:

$\left\{\left(\frac{2}{3}, \frac{2}{3}, 1, \frac{2}{3}, 1\right)\right\}$ is a basis of the eigenspace for the eigenvalue 4.

$\{(2,0,1,0,0),\ (-2,1,0,1,1)\}$ is a basis of the eigenspace for the eigenvalue 3.

$\left\{\left(\frac{1}{2}, 1, 1, 0, 0\right),\ \left(\frac{1}{2}, -1, 0, 1, 1\right)\right\}$ is a basis of the eigenspace for the eigenvalue 2.

We have found five linearly independent eigenvectors for the 5 by 5 matrix A; therefore, we conclude that A is diagonalizable. We use this information to exhibit the diagonalization as follows:

If $D = \begin{bmatrix} 4 & 0 & 0 & 0 & 0 \\ 0 & 3 & 0 & 0 & 0 \\ 0 & 0 & 3 & 0 & 0 \\ 0 & 0 & 0 & 2 & 0 \\ 0 & 0 & 0 & 0 & 2 \end{bmatrix}$ and if $P = \begin{bmatrix} \frac{2}{3} & 2 & -2 & \frac{1}{2} & \frac{1}{2} \\ \frac{2}{3} & 0 & 1 & 1 & -1 \\ 1 & 1 & 0 & 1 & 0 \\ \frac{2}{3} & 0 & 1 & 0 & 1 \\ 1 & 0 & 1 & 0 & 1 \end{bmatrix}$, then $P^{-1}AP = D$.

We remark that our choice of P and D are not the only correct ones. The eigenvalues

```
7:   ROW_REDUCE (4 i - a)

        ⎡                2 ⎤
        ⎢ 1  0  0  0  - ─── ⎥
        ⎢                3 ⎥
        ⎢                  ⎥
        ⎢                2 ⎥
        ⎢ 0  1  0  0  - ─── ⎥
        ⎢                3 ⎥
8:      ⎢                  ⎥
        ⎢ 0  0  1  0    -1 ⎥
        ⎢                  ⎥
        ⎢                2 ⎥
        ⎢ 0  0  0  1  - ─── ⎥
        ⎢                3 ⎥
        ⎢                  ⎥
        ⎣ 0  0  0  0     0 ⎦

9:   ROW_REDUCE (3 i - a)

        ⎡ 1  0  -2  0   2 ⎤
        ⎢ 0  1   0  0  -1 ⎥
10:     ⎢ 0  0   0  1  -1 ⎥
        ⎢ 0  0   0  0   0 ⎥
        ⎣ 0  0   0  0   0 ⎦

11:  ROW_REDUCE (2 i - a)

        ⎡          1          1 ⎤
        ⎢ 1  0  - ───  0  - ─── ⎥
        ⎢          2          2 ⎥
        ⎢                       ⎥
        ⎢ 0  1    -1   0     1  ⎥
12:     ⎢                       ⎥
        ⎢ 0  0     0   1    -1  ⎥
        ⎢                       ⎥
        ⎢ 0  0     0   0     0  ⎥
        ⎢                       ⎥
        ⎣ 0  0     0   0     0  ⎦
```

Figure 15.2: **Eigenvectors of A**

down the diagonal of D may be put in a different order and the columns of P interchanged accordingly. Also, eigenspaces are unique, but there are infinitely many bases for each eigenspace, thus there are infinitely many correct matrices P. To be certain that our diagonalization is correct, we need only check with *DERIVE* that $P^{-1}AP = D$. *When diagonalizing matrices you should always perform this check.*

Solution for B: Enter and name the matrix B and then **Author** , **Simplify** and **Factor** `charpoly(B)` in Figure 15.3 as we did for A. We focus attention on the eigenvalue 2 of order 2 in expression 6. **Author** `row_reduce(2I-B)` and **Simplify**. We see from expression 8 of Figure 15.3 that the null space of this matrix has dimension 1, and hence the eigenspace of B associated with the eigenvalue 2 has dimension 1. Since the eigenvalue 2 has order 2, we conclude without further calculation that B is not a diagonalizable matrix.

```
1                                         2
2:   b := [ -6  -5  12  -9   2 ]         7:   ROW_REDUCE(2 i - b)
          [ -4   1   5  -2  -1 ]
          [ -8  -3  13  -7   0 ]              [ 1  0  0  0  - 3/2 ]
          [ -4  -1   5   0  -1 ]              [ 0  1  0  0   -1   ]
          [ -4  -1   5  -1   0 ]         8:   [ 0  0  1  0   -2   ]
                                              [ 0  0  0  1   -1   ]
3:   i := IDENTITY_MATRIX(5)                  [ 0  0  0  0    0   ]

4:   CHARPOLY(b)

        5       4        3        2
5:   - w  + 8 w  - 23 w  + 28 w  - 12 w

                                  2
6:   w (3 - w) (w - 1) (w - 2)
```

Figure 15.3: **Eigenvalues and eigenvectors of B**

Solved Problem 2: Let $A = \begin{bmatrix} 0 & -\frac{1}{6} & -\frac{1}{6} & -\frac{1}{2} \\ -4 & -\frac{1}{3} & -\frac{4}{3} & -4 \\ -4 & -\frac{4}{3} & -\frac{1}{3} & -4 \\ 3 & \frac{5}{6} & \frac{5}{6} & \frac{7}{2} \end{bmatrix}$.

1. Find A^n for an arbitrary positive integer n.
2. **For students who have studied calculus.** Find $\lim_{n\to\infty} A^n$.
3. Find a cube root of A with real entries.

Solution: The first step is to diagonalize the matrix exactly as we did in Solved Problem 1. We conclude that

$$21: \quad p \cdot \begin{bmatrix} 3^{-n} & 0 & 0 & 0 \\ 0 & 2^{-n} & 0 & 0 \\ 0 & 0 & 1 & 0 \\ 0 & 0 & 0 & 1 \end{bmatrix} \cdot p^{-1}$$

$$22: \quad \begin{bmatrix} 3^{1-n} - 2^{1-n} & 3^{-n} - 2^{-n} & 3^{-n} - 2^{-n} & 3^{1-n} \\ 2\ 3^{1-n} - 6 & 2\ 3^{-n} - 1 & 2\ 3^{-n} - 2 & 2\ 3^{1} \\ 2\ 3^{1-n} - 6 & 2\ 3^{-n} - 2 & 2\ 3^{-n} - 1 & 2\ 3^{1} \\ -2\ 3^{1-n} + 2^{1-n} + 4 & -2\ 3^{-n} + 2^{-n} + 1 & -2\ 3^{-n} + 2^{-n} + 1 & -2\ 3^{1-n} \end{bmatrix}$$

Figure 15.4: **The nth power of a matrix**

$$D = P^{-1}AP, \text{ where } \quad P = \begin{bmatrix} -\frac{1}{2} & -1 & 0 & 0 \\ -1 & 0 & -1 & -3 \\ -1 & 0 & 1 & 0 \\ 1 & 1 & 0 & 1 \end{bmatrix} \quad \text{and } \; D = \begin{bmatrix} \frac{1}{3} & 0 & 0 & 0 \\ 0 & \frac{1}{2} & 0 & 0 \\ 0 & 0 & 1 & 0 \\ 0 & 0 & 0 & 1 \end{bmatrix}$$

To calculate A^n we use the fact that $A^n = (PDP^{-1})^n = PD^nP^{-1}$. (You should be able to prove this, at least for $n = 2$.) Thus, with P and D appropriately defined, we **Author** and **Simplify** `P.D^n.P^-1`. The nth power of A is partially displayed in expression 22 of Figure 15.4.

For part 2, we may use **Calculus Limit**. When we are prompted for the limiting value, type `inf` for infinity. The result is expression 23 in Figure 15.5. **Simplify** it to get the answer in expression 24.

$$23: \lim_{n\to\infty} \begin{bmatrix} 3^{1-n} - 2^{1-n} & 3^{-n} - 2^{-n} & 3^{-n} - 2^{-n} & 3^{1} \\ 2\,3^{1-n} - 6 & 2\,3^{-n} - 1 & 2\,3^{-n} - 2 & 2 \\ 2\,3^{1-n} - 6 & 2\,3^{-n} - 2 & 2\,3^{-n} - 1 & 2 \\ -2\,3^{1-n} + 2^{1-n} + 4 & -2\,3^{-n} + 2^{-n} + 1 & -2\,3^{-n} + 2^{-n} + 1 & -2\,3^{1} \end{bmatrix}$$

$$24: \begin{bmatrix} 0 & 0 & 0 & 0 \\ -6 & -1 & -2 & -6 \\ -6 & -2 & -1 & -6 \\ 4 & 1 & 1 & 4 \end{bmatrix}$$

Figure 15.5: $\boxed{\lim_{n\to\infty} A^n}$

Alternative solution to part 2: Observe that

$$\lim_{n\to\infty} A^n = P(\lim_{n\to\infty} D^n)P^{-1} = P \begin{bmatrix} 0 & 0 & 0 & 0 \\ 0 & 0 & 0 & 0 \\ 0 & 0 & 1 & 0 \\ 0 & 0 & 0 & 1 \end{bmatrix} P^{-1}.$$

Use **Declare Matrix** to enter $\begin{bmatrix} 0 & 0 & 0 & 0 \\ 0 & 0 & 0 & 0 \\ 0 & 0 & 1 & 0 \\ 0 & 0 & 0 & 1 \end{bmatrix}$ and name it Q. **Author** and **Simplify** `P.Q.P^-1` and the result will be the same as before.

For part 3, we want a matrix C with real entries such that $C^3 = A$. Notice that $(PD^{\frac{1}{3}}P^{-1})^3 = P(D^{\frac{1}{3}})^3 P^{-1} = PDP^{-1} = A$. Thus $A^{\frac{1}{3}} = PD^{\frac{1}{3}}P^{-1}$, so we may **Author** and **Simplify** `P.D^(1/3).P^-1`. The result is partially displayed in expression 27 of Figure 15.6. (We will not do it here, but you should check this answer by cubing it to see that it gives the original matrix: **Author** and **Simplify** `(#27)^3`.)

It may occur to you to use **Manage Substitute** to substitute $\frac{1}{3}$ for n in expression 22 of Figure 15.4. This works, but it has to be justified because the formula was only shown to be valid for positive integers. The argument above presents the justification.

$$27: \quad \begin{bmatrix} 3^{2/3} - 2^{2/3} & \frac{3^{2/3}}{3} - \frac{2^{2/3}}{2} & \frac{3^{2/3}}{3} - \frac{2^{2/3}}{2} & 3^{2/3} \\ 2\,3^{2/3} - 6 & \frac{2\,3^{2/3}}{3} - 1 & \frac{2\,3^{2/3}}{3} - 2 & 2 \\ 2\,3^{2/3} - 6 & \frac{2\,3^{2/3}}{3} - 2 & \frac{2\,3^{2/3}}{3} - 1 & 2 \\ 2^{2/3} - 2\,3^{2/3} + 4 & \frac{2^{2/3}}{2} - \frac{2\,3^{2/3}}{3} + 1 & \frac{2^{2/3}}{2} - \frac{2\,3^{2/3}}{3} + 1 & \frac{3\,2^{2/3}}{2} \end{bmatrix}$$

Figure 15.6: **A cube root of a matrix**

Solved Problem 3: Orthogonally diagonalize the symmetric matrix $A = \begin{bmatrix} 3 & -1 & -1 & 1 \\ -1 & 3 & 1 & -1 \\ -1 & 1 & 3 & -1 \\ 1 & -1 & -1 & 3 \end{bmatrix}$.

Also, find a matrix B such that $BB^t = A$.

Solution: To orthogonally diagonalize a symmetric matrix, we have to find an *orthogonal* matrix Q such that $Q^{-1}AQ$ is diagonal. (Recall that when Q is orthogonal, $Q^{-1} = Q^t$.) Follow the procedure in Solved Problem 1 to find the eigenvalues and eigenvectors for A. From Figure 15.7 we obtain the following:

$\{(1, -1, -1, 1)\}$ is a basis of the eigenspace for the eigenvalue 6.

$\{(1, 1, 0, 0),\ (1, 0, 1, 0),\ (-1, 0, 0, 1)\}$ is a basis of the eigenspace for the eigenvalue 2.

Remark: Notice that the basis vector for eigenvalue 6 is orthogonal to all the basis vectors for eigenvalue 2. When A is symmetric, vectors in eigenspaces corresponding to distinct eigenvalues are orthogonal.

```
1:                                          2:

                                            7:   ROW_REDUCE (6 i - a)

         ┌  3  -1  -1   1 ┐                      ┌ 1  0  0  -1 ┐
         │ -1   3   1  -1 │                      │ 0  1  0   1 │
2:  a := │ -1   1   3  -1 │                 8:   │ 0  0  1   1 │
         └  1  -1  -1   3 ┘                      └ 0  0  0   0 ┘

3:  i := IDENTITY_MATRIX (4)                9:   ROW_REDUCE (2 i - a)

4:  CHARPOLY (a)                                 ┌ 1  -1  -1  1 ┐
                                                 │ 0   0   0  0 │
     4       3       2                      10:  │ 0   0   0  0 │
5:  w  - 12 w  + 48 w  - 80 w + 48               └ 0   0   0  0 ┘

                      3
6:  (w - 6) (w - 2)
```

Figure 15.7: **Eigenvalues and eigenvectors for a symmetric matrix**

Now we have to convert these bases into orthonormal bases which will then form the columns of the matrix Q. The eigenspace for the eigenvalue 6 has dimension 1, thus we find an orthonormal basis by simply normalizing the single basis vector $(1, -1, -1, 1)$: **Author** and **Simplify** `[1,-1,-1,1]/|[1,-1,-1,1]|`. The orthonormal basis appears in expression 2 of Figure 15.8.

The eigenspace for the eigenvalue 2 has dimension 3 and we must apply the Gram-Schmidt process to obtain an orthonormal basis. This is seen in expressions 4 through 9 of Figure 15.8.

The desired matrix, Q, should have as its columns the vector in expression 2 and the normalized vectors `v1, v2, v3`. This is accomplished when we **Author**

```
[#2, v1/|v1|, v2/|v2|, v3/|v3|]'
```

The result is seen as expression 10 in Figure 15.8. This has been simplified and named Q in expression 12 of Figure 15.9.

```
2:    [1/2, -1/2, -1/2, 1/2]

               u · v
3:   P(u, v) := ——————— v
               v · v

4:   u1 := [1, 1, 0, 0]

5:   u2 := [1, 0, 1, 0]

6:   u3 := [-1, 0, 0, 1]

7:   v1 := u1

8:   v2 := u2 - P(u2, v1)

9:   v3 := u3 - P(u3, v1) - P(u3, v2)

10:  [[1/2, -1/2, -1/2, 1/2], v1/|v1|, v2/|v2|, v3/|v3|]
```

Figure 15.8: **Finding an orthonormal basis**

We will not do so here, but you should use *DERIVE* to check that $Q^tQ = I$ and that Q^tAQ is the diagonal matrix $D = \begin{bmatrix} 6 & 0 & 0 & 0 \\ 0 & 2 & 0 & 0 \\ 0 & 0 & 2 & 0 \\ 0 & 0 & 0 & 2 \end{bmatrix}$.

To solve the second part of the problem, observe first that if $E = \begin{bmatrix} \sqrt{6} & 0 & 0 & 0 \\ 0 & \sqrt{2} & 0 & 0 \\ 0 & 0 & \sqrt{2} & 0 \\ 0 & 0 & 0 & \sqrt{2} \end{bmatrix}$, then $EE^t = E^2 = D$. Thus $A = QDQ^t = Q(EE^t)Q^t = (QE)(QE)^t$. The solution $B = QE$ is expression 15 of Figure 15.9. (Once again, you should use *DERIVE* to check that $B^tB = A$.)

1 ═══ 2 ═══

$$12:\ \blacksquare := \begin{bmatrix} \frac{1}{2} & \frac{\sqrt{2}}{2} & \frac{\sqrt{6}}{6} & -\frac{\sqrt{3}}{6} \\ -\frac{1}{2} & \frac{\sqrt{2}}{2} & -\frac{\sqrt{6}}{6} & \frac{\sqrt{3}}{6} \\ -\frac{1}{2} & 0 & \frac{\sqrt{6}}{3} & \frac{\sqrt{3}}{6} \\ \frac{1}{2} & 0 & 0 & \frac{\sqrt{3}}{2} \end{bmatrix}$$

$$14:\ \blacksquare \cdot \begin{bmatrix} \sqrt{6} & 0 & 0 & 0 \\ 0 & \sqrt{2} & 0 & 0 \\ 0 & 0 & \sqrt{2} & 0 \\ 0 & 0 & 0 & \sqrt{2} \end{bmatrix}$$

$$15:\ \begin{bmatrix} \frac{\sqrt{6}}{2} & 1 & \frac{\sqrt{3}}{3} & -\frac{\sqrt{6}}{6} \\ -\frac{\sqrt{6}}{2} & 1 & -\frac{\sqrt{3}}{3} & \frac{\sqrt{6}}{6} \\ -\frac{\sqrt{6}}{2} & 0 & \frac{2\sqrt{3}}{3} & \frac{\sqrt{6}}{6} \\ \frac{\sqrt{6}}{2} & 0 & 0 & \frac{\sqrt{6}}{2} \end{bmatrix}$$

Figure 15.9: **Orthogonal diagonalization of A**

Exercises

1. Determine which of the following matrices are diagonalizable. For each that is, exhibit the diagonalization and show all your work.

$$A = \begin{bmatrix} 1 & -3 & -4 & 4 & -2 \\ 10 & 12 & 4 & 0 & 12 \\ 2 & 1 & 1 & 2 & 2 \\ 5 & 5 & 2 & 2 & 6 \\ -9 & -7 & 0 & -4 & -8 \end{bmatrix} \quad B = \begin{bmatrix} 6 & 8 & -12 & 6 & -2 \\ 4 & 4 & -8 & 6 & -2 \\ 8 & 10 & -16 & 9 & -3 \\ 4 & 6 & -8 & 4 & -2 \\ 4 & 6 & -8 & 5 & -3 \end{bmatrix} \quad C = \begin{bmatrix} 16 & 6 & -1 & 7 & 19 \\ -9 & -1 & 1 & -5 & -13 \\ 5 & 2 & 2 & 3 & 7 \\ -7 & -3 & 1 & -1 & -10 \\ -4 & -2 & 0 & -2 & -3 \end{bmatrix}$$

2. Find a cube root of each of the following matrices, with real entries. *Check your final answers!*

$$A = \begin{bmatrix} 4 & 0 & -1 & 1 & 2 \\ -3 & 2 & 3 & 0 & -3 \\ 9 & 0 & -4 & 3 & 6 \\ -5 & 0 & 1 & -2 & -2 \\ 4 & 0 & -2 & 2 & 3 \end{bmatrix} \quad B = \begin{bmatrix} -2 & -4 & 6 & -4 & 2 \\ -2 & -1 & 4 & -4 & 2 \\ -4 & -5 & 9 & -6 & 3 \\ -2 & -3 & 4 & -2 & 2 \\ -2 & -3 & 4 & -4 & 4 \end{bmatrix} \quad C = \begin{bmatrix} 66 & -51 & -12 & 8 & -4 \\ 58 & -43 & -12 & 8 & -4 \\ 96 & -72 & -19 & 12 & -6 \\ 58 & -42 & -12 & 7 & -4 \\ 67 & -51 & -12 & 8 & -5 \end{bmatrix}$$

3. **For students who have studied calculus.** Find the limit as n goes to infinity (if it exists) of the nth power of each of the following matrices.

$$A = \begin{bmatrix} \frac{1}{2} & -\frac{1}{2} & \frac{1}{2} & -\frac{1}{2} & \frac{1}{2} \\ -\frac{1}{2} & -\frac{1}{4} & \frac{5}{4} & -\frac{5}{4} & \frac{1}{2} \\ \frac{1}{6} & -\frac{1}{2} & \frac{5}{6} & -\frac{1}{2} & \frac{1}{2} \\ \frac{2}{3} & 0 & -\frac{2}{3} & 1 & 0 \\ 0 & -\frac{3}{4} & \frac{3}{4} & -\frac{3}{4} & 1 \end{bmatrix} \quad B = \begin{bmatrix} 1 & 9 & -9 & 13 & -13 \\ \frac{1}{2} & -\frac{11}{2} & 6 & -\frac{19}{2} & \frac{19}{2} \\ \frac{1}{2} & 9 & -\frac{17}{2} & 13 & -13 \\ -3 & 0 & 3 & -1 & 0 \\ -3 & -9 & 12 & -15 & 14 \end{bmatrix}$$

4. Orthogonally diagonalize the following matrices.

$$A = \begin{bmatrix} 1 & 0 & 2 & 2 \\ 0 & 1 & 2 & -2 \\ 2 & 2 & 1 & 0 \\ 2 & -2 & 0 & 1 \end{bmatrix} \qquad B = \begin{bmatrix} 4 & 0 & 1 & 1 \\ 0 & 4 & 1 & -1 \\ 1 & 1 & 5 & 0 \\ 1 & -1 & 0 & 5 \end{bmatrix}$$

5. For the matrices of Exercise 4 above, find matrices E and F so that $EE^t = A$ and $FF^t = B$.

6. Orthogonally diagonalize the following matrices.

$$A = \begin{bmatrix} x & -1 & -1 & 1 \\ -1 & x & 1 & -1 \\ -1 & 1 & x & -1 \\ 1 & -1 & -1 & x \end{bmatrix} \qquad B = \begin{bmatrix} 1 & 1 & 1 & x \\ 1 & 1 & x & 1 \\ 1 & x & 1 & 1 \\ x & 1 & 1 & 1 \end{bmatrix}$$

7. A is a 5 by 5 matrix with eigenvalues 1 and 4. A basis for the eigenspace of A associated with 1 is $\{(2,4,3,1,5),(3,5,1,2,2),(1,1,2,4,1)\}$ and a basis for the eigenspace of A associated with 4 is $\{(3,1,3,3,3),(4,2,1,5,2)\}$. Find A.

Exploration and Discovery

1. In Chapter 3 we approximated transcendental functions evaluated at a matrix. We are now in a position to make exact calculations. Let $f(x) = \sum_{n=0}^{\infty} a_n x^n$ be a function given by a power series. If A is a matrix, we define $f(A)$ to be $\sum_{n=0}^{\infty} a_n A^n$, provided that the series converges. Observe that for diagonal matrices D the calculation of $f(D)$ is the same as the evaluation of f at each of the diagonal entries. Thus, for example, $\sin \begin{bmatrix} 1 & 0 & 0 \\ 0 & 2 & 0 \\ 0 & 0 & 3 \end{bmatrix} = \begin{bmatrix} \sin 1 & 0 & 0 \\ 0 & \sin 2 & 0 \\ 0 & 0 & \sin 3 \end{bmatrix}$. Furthermore, if $A = PDP^{-1}$, then we have the following:

$$f(A) = \sum_{n=0}^{\infty} a_n A^n = \sum_{n=0}^{\infty} a_n (PDP^{-1})^n = \sum_{n=0}^{\infty} a_n PD^n P^{-1} = P(\sum_{n=0}^{\infty} a_n D^n)P^{-1} = Pf(D)P^{-1}$$

(a) If $A = \begin{bmatrix} 2 & 1 & -1 \\ 1 & 2 & -1 \\ -1 & 1 & 2 \end{bmatrix}$, $D = \begin{bmatrix} 1 & 0 & 0 \\ 0 & 2 & 0 \\ 0 & 0 & 3 \end{bmatrix}$ and $P = \begin{bmatrix} 1 & 1 & 1 \\ 0 & 1 & 1 \\ 1 & 1 & 0 \end{bmatrix}$, verify that $A = PDP^{-1}$.

(b) Calculate $\sin(A) = P \sin(D) P^{-1}$ and use **approX** to approximate it.

(c) Use Maclaurin polynomials of degrees 9 and 17 for $\sin x$ to approximate $\sin(A)$ as we did in Chapter 3. Compare your results with those in part (b).

(d) Repeat all the above for $\cos x$ and e^x.

(e) For $A = \begin{bmatrix} 16 & 6 & 3 & 15 \\ 12 & 4 & 3 & 15 \\ 36 & 18 & 7 & 45 \\ -24 & -12 & -6 & -32 \end{bmatrix}$ calculate $\sin A$, $\cos A$, and e^A.

2. A matrix is said to be *nilpotent* if some power of A is the zero matrix. Let $A = \begin{bmatrix} 0 & p & q & r \\ 0 & 0 & s & t \\ 0 & 0 & 0 & u \\ 0 & 0 & 0 & 0 \end{bmatrix}$ and let $B = \begin{bmatrix} 0 & p & q & r & s \\ 0 & 0 & t & u & v \\ 0 & 0 & 0 & w & x \\ 0 & 0 & 0 & 0 & y \\ 0 & 0 & 0 & 0 & 0 \end{bmatrix}$. Show that A and B are nilpotent. What can you say about a matrix C all of whose entries on the main diagonal and below are zero? Which nilpotent matrices are diagonalizable?

LABORATORY EXERCISE 15.1

Diagonalizing a Matrix

Name ______________________________ **Due Date** ______________

Let $A = \begin{bmatrix} 1 & -2 & 4 & 2 \\ -2 & 1 & 4 & 2 \\ 0 & -2 & 5 & 2 \\ -2 & -2 & 4 & 5 \end{bmatrix}$.

1. Find the characteristic polynomial of A.

2. Find the exact eigenvalues of A.

3. Find a basis for each eigenspace of A.

4. Find a matrix P such that $P^{-1}AP$ is diagonal.

5. Let D be the diagonal matrix in Part 4 above. Find AD. How is AD related to P? Explain.

CHAPTER 16

MATRICES AND LINEAR TRANSFORMATIONS FROM R^n TO R^m

LINEAR ALGEBRA CONCEPTS

- **Linear transformation**
- **Matrix of a linear transformation**
- **Kernel of a linear transformation**
- **Image of a linear transformation**
- **Inverse of a linear transformation**
- **Composition of linear transformations**

Introduction

Geometrically, linear transformations between vector spaces preserve straight lines. Algebraically, they preserve addition and scalar multiplication, the linear structure of the space. A linear transformation from R^m to R^n has a particularly simple description as a matrix product.

We will use the notation $[T]$ for the matrix of T with respect to the standard basis and call it *the standard matrix of* T. It is defined as the n by m matrix whose ith column is $T(\mathbf{e_i})$, where $\{\mathbf{e_1}, \mathbf{e_2}, \ldots, \mathbf{e_m}\}$ denotes the standard basis for R^m.

The following theorem summarizes many of the standard matrix and will be useful throughout this chapter.

Theorem: Let $T : R^m \rightarrow R^n$ be a linear transformation.

1. For each vector $\mathbf{v}$ in R^m, $T(\mathbf{v}) = [T]\mathbf{v}$.
2. The kernel of T is the null space of $[T]$.
3. The image of T is the column space of $[T]$.
4. If $m = n$, then the standard matrix of T^k is $[T]^k$.
5. If $m = n$, then T has an inverse if and only if $[T]$ is invertible, in which case the standard matrix of T^{-1} is $[T]^{-1}$.

Solved Problems

Solved Problem 1: Let $T: R^3 \to R^4$ be defined by

$$T(x,y,z) = (x+4y-2z, 2x+y-z, 3x+5y-3z, -x+3y-z)$$

and let $S: R^2 \to R^1$ be defined by $S(x,y) = \text{DET}\begin{bmatrix} x & y \\ x+y & x-y \end{bmatrix}$. Determine if T and S are linear transformations.

Solution: **Author** `T(x,y,z):=[x+4y-2z,2x+y-z,3x+5y-3z,-x+3y-z]`. T is a linear transformation provided that it satisfies the following two conditions for all x, y, z, a, b, c.

$$\begin{aligned} T((x,y,z)+(a,b,c)) &= T(x,y,z)+T(a,b,c) \\ T(c(x,y,z)) &= cT(x,y,z) \end{aligned}$$

or equivalently,

$$\begin{aligned} T(x+a, y+b, z+c) - T(x,y,z) - T(a,b,c) &= 0 \\ T(cx, cy, cz) - cT(x,y,z) &= 0 \end{aligned}$$

Thus we **Author** and **Simplify** `T(x+a,y+b,z+c)-T(x,y,z)-T(a,b,c)` and `T(cx,cy,cz)-cT(x,y,z)`. From expressions 2 to 5 of Figure 16.1 we see that both simplify to the zero vector, and we conclude that T is a linear transformation.

The second part of the exercise is easy to do by hand and certainly does not require *DERIVE*, but we present this nonlinear case for comparison.

Author `S(x,y):=det[[x,y],[x+y,x-y]]`. Proceeding as above, **Author** and **Simplify** `S(x+a,y+b)-S(x,y)-S(a,b)`. The result displayed in expression 8 of Figure 16.1 does not appear to be zero, but to verify it you should select specific values of x, y, a, b that do not yield 0. Choosing $x = y = a = b = 1$ gives $S(1+1, 1+1) - S(1,1) - S(1,1) = -2 \neq 0$, so S is not a linear transformation.

```
1:   T (x, y, z) := [x + 4 y - 2 z, 2 x + y - z, 3 x + 5 y - 3 z, -x + 3 y - z]

2:   T (x + a, y + b, z + c) - T (x, y, z) - T (a, b, c)

3:   [0, 0, 0, 0]

4:   T (c x, c y, c z) - c T (x, y, z)

5:   [0, 0, 0, 0]

                      ┌  x       y   ┐
6:   S (x, y) := DET  │              │
                      └ x + y  x - y ┘

7:   S (x + a, y + b) - S (x, y) - S (a, b)

8:   x (2 a - 2 b) - 2 y (a + b)
```

Figure 16.1: **Testing functions to determine if they are linear transformations**

<u>**Solved Problem 2**</u>: Let $T : R^2 \to R^2$ be defined by $T(x, y) = (x + 4y, 2x + y)$ and let $S : R^2 \to R^2$ be defined by $S(x, y) = (xy, 2x + y)$.

Examine the images under T and S of two lines as a test for linearity.

<u>Solution</u>: **Author** `[x+4y,2x+y]` as seen in Figure 16.2. Let's consider the line $y = x$. Use **Manage Substitute** to replace x by y. When it's simplified we get expression 3. Now **Plot Beside**. Press Enter to accept the parameter interval -3.14159 and 3.14159. When we **Plot** this we see a straight line. If we repeat this for $x = 2y$ in expressions 5 and 6 of Figure 16.2, we see a second line. Thus, we suspect T is linear because it transforms lines to lines. (To *prove* it's linear we must verify the definition $T(a\mathbf{x} + \mathbf{y}) = aT(\mathbf{x}) + T(\mathbf{y})$.)

Author `[xy,2x+y]` as seen in expression 6 of Figure 16.3. Let's consider the line $y = x$. Use **Manage Substitute** to replace x by y. Now **Plot Beside**. Press Enter to accept the parameter interval -3.14159 and 3.14159. When we plot this we see a curve, actually a parabola. If we repeat this for $x = 2y$, we see a second parabola. This shows that S transforms straight lines into curves that are not lines. Thus, S is nonlinear.

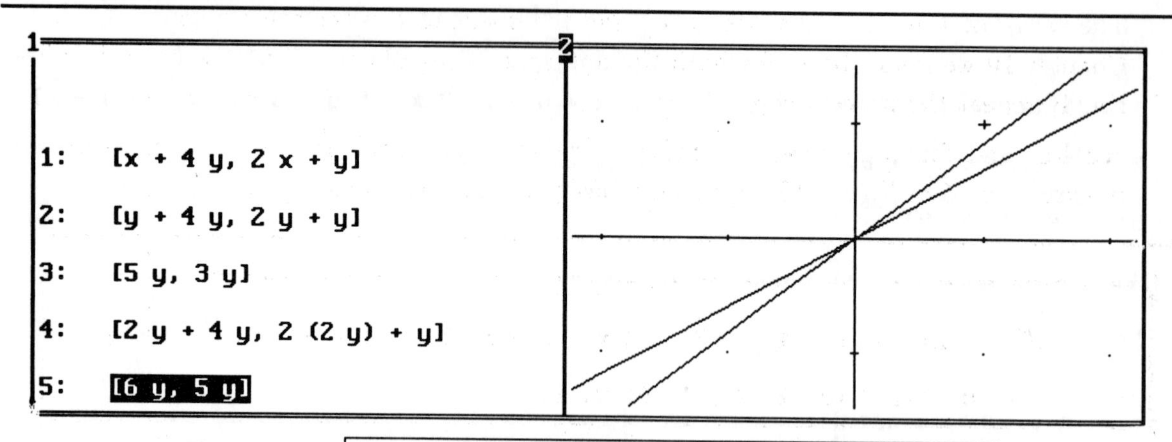

Figure 16.2: Image of lines under a linear transformation

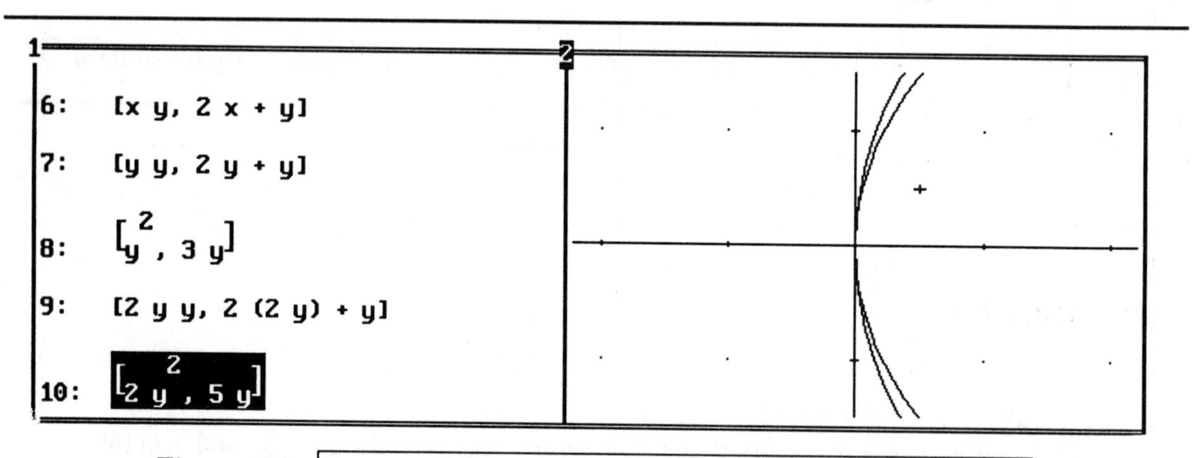

Figure 16.3: Image of lines under a non-linear transformation

<u>**Solved Problem 3**</u>**:** Let T be the linear transformation from Solved Problem 1. Find the standard matrix of T and find bases for the image and kernel of T.

<u>Solution</u>: To get the standard matrix of T it's probably most efficient to simply jot it down on paper and then use **Declare Matrix** to enter it. However, we can also **Author** `T(x,y,z):=[x+4y-2z,2x+y-z,3x+5y-3z,-x+3y-z]` and then **Author** and **Simplify** `[T(1,0,0),T(0,1,0),T(0,0,1)]'` as seen in Figure 16.4. (Notice the transpose symbol (') at the end.) The standard matrix $[T]$ of T appears in expression 3 of Figure 16.4.

The kernel of T is the null space of $[T]$, and the image of T is the column space of $[T]$. In Chapter 10 we found bases for both the null space and column space of a matrix, so we simply repeat the process here.

Author and **Simplify** `row_reduce(#3)`. From the reduced form of the matrix displayed in expression 5 of Figure 16.4 we read a basis for the kernel of T: $\{(\frac{2}{7}, \frac{3}{7}, 1)\}$.

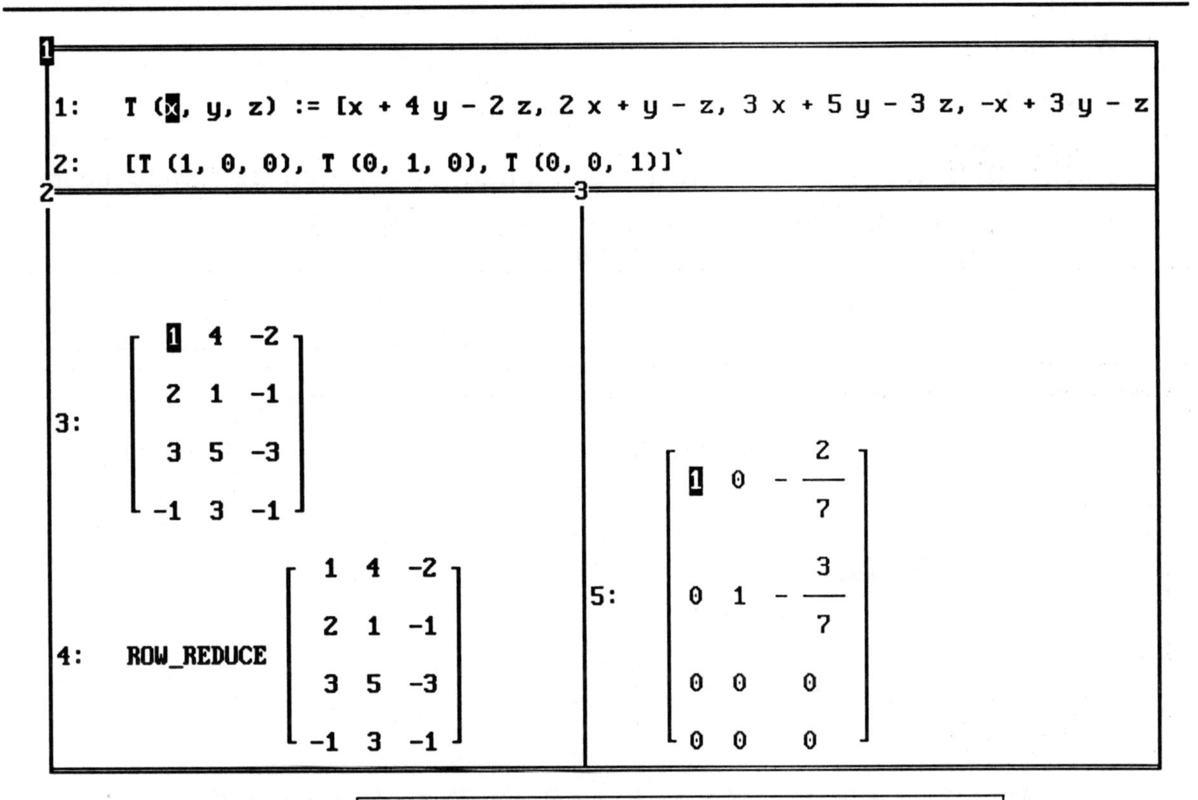

Figure 16.4: **Kernel and image of a linear transformation**

From expression 5 of Figure 16.4 the first two columns of the reduced form of $[T]$ contain leading ones. Thus the first two columns of $[T]$ form a basis for the image of T: $\{(1, 2, 3, -1), (4, 1, 5, 3)\}$.

<u>**Solved Problem 4**</u>**:** Let $T : R^3 \to R^3$ be defined by
$T(x, y, z) = (x - 2y + 3z, 2x + y - z, x + 3y + 2z)$.

1. Find the standard matrix of T.
2. Find $T^3(x, y, z)$.
3. Show that T has an inverse and find $T^{-1}(x, y, z)$.

Solution: As in Solved Problem 2, we can just jot down the standard matrix on paper and then use **Declare Matrix** to enter it, or we can **Author** `T(x,y,z):=[x-2y+3z,2x+y-z,x+3y+2z]` and then **Author** and **Simplify** `[T(1,0,0),T(0,1,0),T(0,0,1)]'`. We did it the first way to get expression 1 in Figure 16.5. Since we will need the vector (x, y, z), we **Author** it as in expression 2.

$T^3(\mathbf{x}) = [T]^3\mathbf{x}$, so we **Author** and **Simplify** `(#1)^3 .#2`. (Note the space after the 3.) We conclude from expression 4 in Figure 16.5 that $T^3(x, y, z) = [21x + 38y + 17z, -6x - 3y + 21z, 27x + y + 28z]$.

The linear transformation T has an inverse if and only if the matrix $[T]$ does, in which case $T^{-1}(\mathbf{x}) = [T]^{-1}\mathbf{x}$. Therefore, we **Author** and **Simplify** `(#1)^-1 . #2`. (Note the space after the 1.) We conclude from expression 6 in Figure 16.5 that the matrix $[T]$ has an inverse and that $T^{-1}(x, y, z) = [x/6+13y/30-z/30,\ -x/6-y/30+7z/30,\ x/6-y/6+z/6]$. (If the matrix $[T]$ did not have an inverse, expression 6 would be the same as 5.)

1: $$\begin{bmatrix} 1 & -2 & 3 \\ 2 & 1 & -1 \\ 1 & 3 & 2 \end{bmatrix}$$

2: [x, y, z]

3: $$\begin{bmatrix} 1 & -2 & 3 \\ 2 & 1 & -1 \\ 1 & 3 & 2 \end{bmatrix}^3 \cdot [x, y, z]$$

4: [21 x + 38 y + 17 z, - 6 x - 3 y + 21 z, 27 x + y + 28 z]

5: $$\begin{bmatrix} 1 & -2 & 3 \\ 2 & 1 & -1 \\ 1 & 3 & 2 \end{bmatrix}^{-1} \cdot [x, y, z]$$

6: $$\left[\frac{x}{6} + \frac{13\,y}{30} - \frac{z}{30},\ -\frac{x}{6} - \frac{y}{30} + \frac{7\,z}{30},\ \frac{x}{6} - \frac{y}{6} + \frac{z}{6}\right]$$

Figure 16.5: **Inverse and power of a linear transformation**

Exercises

1. Determine if the following are linear transformations.

   (a) $T : R^3 \to R^3$, $T(x, y, z) := (3x - 5y + 4z, 8x - 7y - 4z, 2x + 5y - \pi z)$

   (b) $T : R^3 \to R^3$, $T(x, y, z) := ((x - y)^2, (x - z)^2, (y - z)^2)$

   (c) $T : M_{2,2} \to M_{2,2}$, $T(A) := A^2 - 3A$

2. Repeat Solved Problem 2 for $T(x, y) = (x^2, y^2)$ and for $S(x, y) = (x \sin^2 y, x \cos^2 y)$. Discuss your observations and conclusions. Are T and S linear?

3. For each of the following linear transformations T, find the standard matrix of T and find bases for both the image and kernel of T.

   (a) $T : R^4 \to R^3$, $T(x, y, z, w) = (2x - 3y + 4z - w, 5x - y + 2z + 4w, 3x + 2y - 2z + 5w)$

   (b) $T : R^4 \to R^3$, $T(\mathbf{v}) = A\mathbf{v}$, where $A = \begin{bmatrix} 2 & 3 & 1 & 1 \\ 3 & 2 & 3 & 3 \\ 5 & 5 & 4 & 4 \end{bmatrix}$

4. For each of the following linear transformations T, find $T^5(x, y, z, w)$. Determine if T^{-1} exists. If it does, find $T^{-1}(x, y, z, w)$.

   (a) $T : R^4 \to R^4$, $T(x, y, z, w) = (x - y - z, x + 2y + 3w, x + 4y + z + w, 3x - 2y + z - 2w)$.

   (b) $T : R^4 \to R^4$, $T(\mathbf{v}) = \begin{bmatrix} 1 & 1 & 2 & 4 \\ 2 & 1 & 3 & 2 \\ 4 & 2 & 1 & 3 \\ 7 & 2 & 3 & 1 \end{bmatrix} \mathbf{v}$

LABORATORY EXERCISE 16.1

Matrices and Linear Transformations

Name ______________________________ **Due Date** ______________

Let $T : R^4 \rightarrow R^4$, $T(x, y, z, w) = (2x + z - w, -y + z + w, x + y - 2w, x - 2w)$.

1. Find the standard matrix of T.

2. Find a basis for the image of T.

3. Find a basis for the kernel of T.

4. Determine if T^{-1} exists. If it does, find $T^{-1}(x, y, z, w)$.

CHAPTER 17

MATRICES OF GENERAL LINEAR TRANSFORMATIONS; SIMILARITY

LINEAR ALGEBRA CONCEPTS

- **Matrix of a linear transformation**
- **Similar matrices**

Introduction

A linear transformation between finite-dimensional vector spaces can be represented by matrix multiplication once ordered bases have been selected for the two vector spaces. A given linear transformation can be represented by different matrices depending on the chosen bases.

Solved Problems

Solved Problem 1: Let $T : R^3 \to R^4$ be the linear transformation defined by $T(x, y, z) = (2x + 3y - 5z, 3x + y + 4z, 2x - 3y + z, x - y + 2z)$, let $B = \{(1,1,2), (2,1,1), (3,0,1)\}$ and let $B' = \{(2,1,1,1), (1,3,2,1), (2,1,1,2), (1,3,3,2)\}$. Find the matrix of T with respect to the bases B and B'.

Solution: Recall that the columns of A are the B' coordinates of the images of elements of B. Thus if $[T]$ is the standard matrix of T, M is the matrix whose columns are the vectors in B and N is the matrix whose columns are the vectors in B', then $A = N^{-1}[T]M$.

The easiest way to enter the standard matrix is to calculate it by hand and use **Declare Matrix**. Alternatively, you may **Author**

```
T(x,y,z):=[2x+3y-5z,3x+y+4z,2x-3y+z,x-y+2z]
[T(1,0,0),T(0,1,0), T(0,0,1)]'
```

and

as separate expressions in Figure 17.1. The standard matrix of T appears in expression 3.

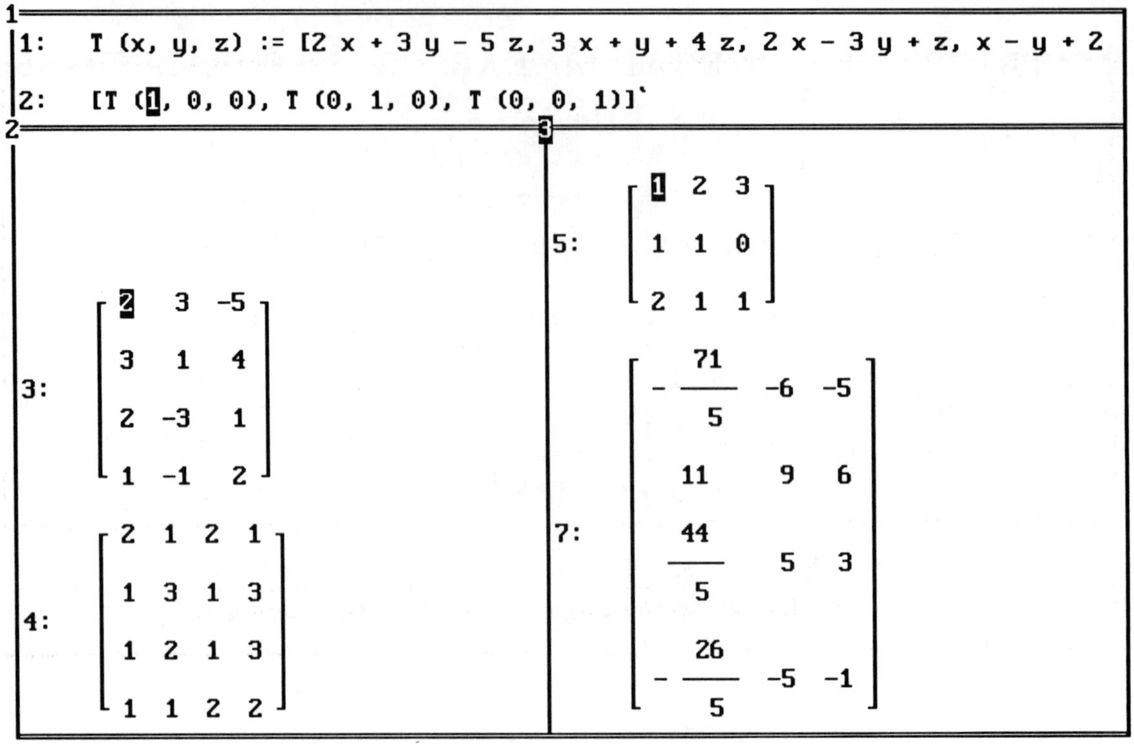

```
1:  T (x, y, z) := [2 x + 3 y - 5 z, 3 x + y + 4 z, 2 x - 3 y + z, x - y + 2
2:  [T (1, 0, 0), T (0, 1, 0), T (0, 0, 1)]`
```

3: $\begin{bmatrix} 2 & 3 & -5 \\ 3 & 1 & 4 \\ 2 & -3 & 1 \\ 1 & -1 & 2 \end{bmatrix}$

4: $\begin{bmatrix} 2 & 1 & 2 & 1 \\ 1 & 3 & 1 & 3 \\ 1 & 2 & 1 & 3 \\ 1 & 1 & 2 & 2 \end{bmatrix}$

5: $\begin{bmatrix} 1 & 2 & 3 \\ 1 & 1 & 0 \\ 2 & 1 & 1 \end{bmatrix}$

7: $\begin{bmatrix} -\frac{71}{5} & -6 & -5 \\ 11 & 9 & 6 \\ \frac{44}{5} & 5 & 3 \\ -\frac{26}{5} & -5 & -1 \end{bmatrix}$

Figure 17.1: **A nonstandard matrix of a linear transformation**

Next use **Declare Matrix** to enter $\begin{bmatrix} 2 & 1 & 2 & 1 \\ 1 & 3 & 1 & 3 \\ 1 & 2 & 1 & 3 \\ 1 & 1 & 2 & 2 \end{bmatrix}$ and $\begin{bmatrix} 1 & 2 & 3 \\ 1 & 1 & 0 \\ 2 & 1 & 1 \end{bmatrix}$. These two matrices appear in expressions 4 and 5 of Figure 17.1. Finally **Author** and **Simplify** `(#4)^-1 .#3 .#5`. The matrix of T with respect to B and B' is expression 7 of Figure 17.1. (Note that expression 6 has been deleted from this Figure to save space.)

Solved Problem 2: Let $T : P_3 \to M_{2,2}$ be the linear transformation defined by

$$T(p) = \begin{bmatrix} p(1) & p(2) \\ p(1)+p(2) & p(1)-p(2) \end{bmatrix}.$$

1. Find the standard matrix of T.
2. Find a basis for the kernel of T.
3. Find a basis for the image of T.

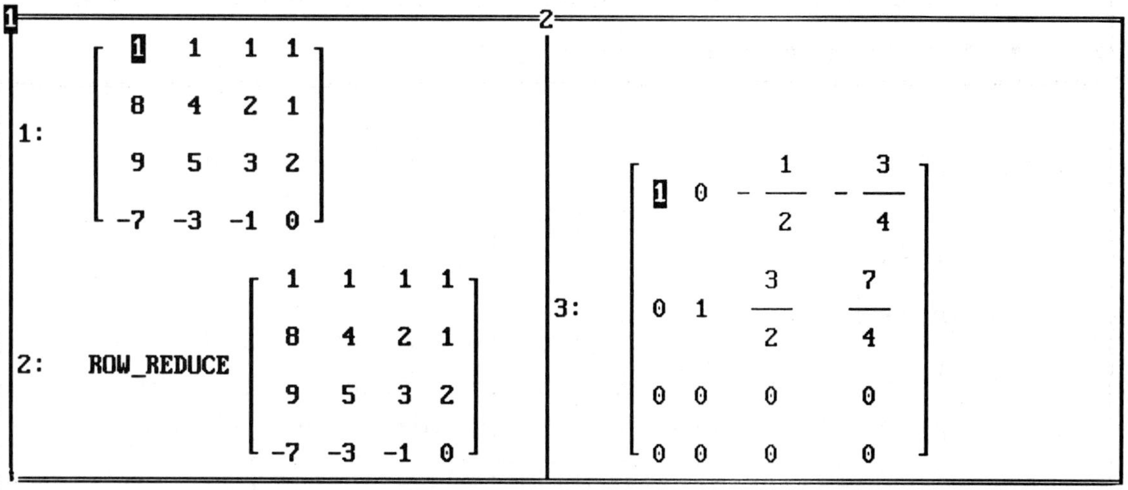

1: $\begin{bmatrix} 1 & 1 & 1 & 1 \\ 8 & 4 & 2 & 1 \\ 9 & 5 & 3 & 2 \\ -7 & -3 & -1 & 0 \end{bmatrix}$

2: ROW_REDUCE $\begin{bmatrix} 1 & 1 & 1 & 1 \\ 8 & 4 & 2 & 1 \\ 9 & 5 & 3 & 2 \\ -7 & -3 & -1 & 0 \end{bmatrix}$

3: $\begin{bmatrix} 1 & 0 & -\frac{1}{2} & -\frac{3}{4} \\ 0 & 1 & \frac{3}{2} & \frac{7}{4} \\ 0 & 0 & 0 & 0 \\ 0 & 0 & 0 & 0 \end{bmatrix}$

Figure 17.2: **Kernel and image of a linear transformation**

Solution: Recall that the standard basis for P_3 is $B = \left\{x^3, x^2, x, 1\right\}$ and that the standard basis for $M_{2,2}$ is

$$B' = \left\{ \begin{bmatrix} 1 & 0 \\ 0 & 0 \end{bmatrix}, \begin{bmatrix} 0 & 1 \\ 0 & 0 \end{bmatrix}, \begin{bmatrix} 0 & 0 \\ 1 & 0 \end{bmatrix}, \begin{bmatrix} 0 & 0 \\ 0 & 1 \end{bmatrix} \right\}$$

To find the standard matrix, we need the B' coordinates of

$$T(x^3) = \begin{bmatrix} 1 & 8 \\ 9 & -7 \end{bmatrix}, \ T(x^2) = \begin{bmatrix} 1 & 4 \\ 5 & -3 \end{bmatrix}, \ T(x) = \begin{bmatrix} 1 & 2 \\ 3 & -1 \end{bmatrix}, \quad \text{and } T(1) = \begin{bmatrix} 1 & 1 \\ 2 & 0 \end{bmatrix}$$

The B' coordinates of these vectors go into the columns of the standard matrix,

$$A = \begin{bmatrix} 1 & 1 & 1 & 1 \\ 8 & 4 & 2 & 1 \\ 9 & 5 & 3 & 2 \\ -7 & -3 & -1 & 0 \end{bmatrix}.$$

Enter this matrix using **Declare Matrix**; then row-reduce it using the ROW_REDUCE function. From expression 3 of Figure 17.2 we read a basis for the null space of A as $\left\{(\frac{1}{2}, -\frac{3}{2}, 1, 0), (\frac{3}{4}, -\frac{7}{4}, 0, 1)\right\}$. These are the B coordinates of a basis for the kernel of T. Thus a basis for the kernel of T is $\left\{\frac{1}{2}x^3 - \frac{3}{2}x^2 + x, \ \frac{3}{4}x^3 - \frac{7}{4}x^2 + 1\right\}$.

Similarly, a basis for the column space of A is $\{(1,8,9,-7),(1,4,5,-3)\}$. These are the B' coordinates of a basis for the image of T. Thus a basis for the image of T is $\left\{\begin{bmatrix} 1 & 8 \\ 9 & -7 \end{bmatrix}, \begin{bmatrix} 1 & 4 \\ 5 & -3 \end{bmatrix}\right\}$.

<u>**Solved Problem 3**</u>: Let $A = \begin{bmatrix} 4 & 1 & -1 \\ -2 & 5 & 2 \\ 4 & -4 & 1 \end{bmatrix}$, $B = \begin{bmatrix} 3 & 3 & 1 \\ -1 & -1 & -1 \\ 1 & 10 & 6 \end{bmatrix}$ and $C = \begin{bmatrix} 3 & 1 & 0 \\ -1 & 3 & 1 \\ 1 & 1 & 2 \end{bmatrix}$.

1. Determine if A and B are similar.
2. Determine if B and C are similar.

<u>Solution</u>: Using **Declare Matrix**, enter the matrices A and B and name them as seen in Figure 17.3. Enter P as a matrix with variable entries as seen in expression 3 of Figure 17.3. *(Be careful not to use a, b, or c as variables.)*

```
1:  a := [  4   1  -1 ]
         [ -2   5   2 ]
         [  4  -4   1 ]

2:  b := [  3   3   1 ]
         [ -1  -1  -1 ]
         [  1  10   6 ]

3:  p := [ d  e  f ]
         [ g  h  i ]
         [ j  k  l ]

4:  APPEND (p · a = b · p)

5:  [4 d - 2 e + 4 f = 3 d + 3 g + j, d + 5 e - 4 f = 3 e + 3 h + k, -d + 2 e +

6:  [d = 0, e = 0, f = 0, g = 0, h = 0, i = 0, j = 0, k = 0, l = 0]
```

Figure 17.3: Matrices that are not similar

We seek an invertible matrix P such that $PAP^{-1} = B$. This matrix equation is too difficult for *DERIVE* to deal with, so we multiply both sides of the equation on the right by P to get $PA = BP$. (Notice that, unlike the first equation, this one may have noninvertible solutions, the zero matrix, for example.)

Author and **Simplify** `append(P.A=B.P)` as seen in expressions 4 and 5 of Figure 17.3. (Recall that APPEND changes a matrix of equations into the form that *DERIVE* can solve.) If we **soLve** the resulting system of equations, we see from expression 6 of Figure 17.3 that the only solution of the equation $PA = BP$ is the zero matrix. Since the zero matrix is not invertible, we conclude that the equation $PAP^{-1} = B$ has no solution and hence the matrices A and B are not similar.

Remark: There is an easier way to solve this problem. Recall that similar matrices must have the same determinant. *DERIVE* can verify that the determinants of A and B are different and hence the matrices are not similar. However, if the determinants were the same, we would have to go through this procedure.

To solve the second part of the problem, enter the matrix C and name it as in Figure 17.4. As in part 1, **Author** `append(P.B=C.P)`; then **Simplify** and **soLve** . From expression 10 of Figure 17.4 we see that the equation $PB = CP$ has infinitely many solutions. We need only choose one that yields an invertible matrix P. In expression 11, we have used **Manage Substitute** to set $@1 = 1, @2 = 2$ and $@3 = 3$. (There are many other choices that will lead to a correct answer.) We **Simplify** expression 11 to get expression 12 of Figure 17.4. From this we see that this choice of parameters leads to the solution $P = \begin{bmatrix} 1 & 2 & 3 \\ 1 & 25 & 8 \\ -16 & -15 & 3 \end{bmatrix}$. Notice that in expressions 13 and 14 of Figure 17.4 we have checked to see that the determinant of P is not zero, thus ensuring that P has an inverse. We conclude that C is similar to B.

Solved Problem 4: Let $T(x, y, z, w) = (5x+y+z+3w, 4x+3y+2z+4w, 2x+4z+2w, -4x-y-2z-2w)$. Find a basis B for R^4 so that the matrix of T with respect to the basis B is a diagonal matrix.

Solution: The matrix of T with respect to the basis B is diagonal if and only if B consists of eigenvectors of T. The procedure for obtaining eigenvectors from the standard matrix of T should be familiar. The standard matrix of T is $A = \begin{bmatrix} 5 & 1 & 1 & 3 \\ 4 & 3 & 2 & 4 \\ 2 & 0 & 4 & 2 \\ -4 & -1 & -2 & -2 \end{bmatrix}$. Enter this matrix and name it A. To find the eigenvalues of A, **Author** `charpoly(A)` and then **Simplify** and **Factor** it. The result is expression 4 of Figure 17.5, from which we see that the eigenvalues of A are 3 and 2.

```
                 [ 3  1  0 ]
7:    c :=       [ -1 3  1 ]
                 [ 1  1  2 ]

8:    APPEND (p · b = c · p)

9:    [3 d - e + f = 3 d + g, 3 d - e + 10 f = 3 e + h, d - e + 6 f = 3 f + i, 3

10:   [d = @1, e = @2, f = @3, g = @3 - @2, h = 3 @1 - 4 @2 + 10 @3, i = @1 - @2

11:   [d = 1, e = 2, f = 3, g = 3 - 2, h = 3 1 - 4 2 + 10 3, i = 1 - 2 + 3 3, j =

12:   [d = 1, e = 2, f = 3, g = 1, h = 25, i = 8, j = -16, k = -15, l = 3]

           [ 1     2    3 ]
13:   DET  [ 1     25   8 ]
           [ -16  -15   3 ]

14:   1088
```

Figure 17.4: **Matrices that are similar**

To get the eigenvectors belonging to 3, **Author** and **Simplify** `row_reduce(3 identity_matrix(4)-A)`. From the result in expression 6, we read a basis for the eigenspace corresponding to 3 as $\left\{(-\frac{1}{2},0,1,0),(-1,-1,0,1)\right\}$.
Similarly, for the eigenvalue 2 we read from expression 8 the basis $\{(-1,2,1,0),(-1,0,0,1)\}$.
Therefore, $B=\left\{(-\frac{1}{2},0,1,0),(-1,-1,0,1),(-1,2,1,0),(-1,0,0,1)\right\}$.

```
1══════════════════════════════════════2══════════════════════════════

2:   CHARPOLY (a)

      4        3       2
3:   w  - 10 w  + 37 w  - 60 w + 36

            2        2
4:   (w - 3)  (w - 2)

5:   ROW_REDUCE (3 IDENTITY_MATRIX (4) - a)
                                              7:   ROW_REDUCE (2 IDENTITY_MATRIX
         ┌            1     ┐
         │ 1   0    ───   1 │                      ┌ 1   0   1   1 ┐
         │            2     │                      │               │
6:       │ 0   1    0     1 │                      │ 0   1  -2   0 │
         │                  │                 8:   │               │
         │ 0   0    0     0 │                      │ 0   0   0   0 │
         │                  │                      │               │
         └ 0   0    0     0 ┘                      └ 0   0   0   0 ┘
```

Figure 17.5: **Diagonalizing a linear transformation**

Exercises

1. Let $T : R^4 \to R^3$ be the linear transformation defined by $T(x,y,z,w) = (x+y-z+w, x+2y+3z, 5x+4y-2w)$. Let $B = \{(2,1,3,3),(3,2,1,4),(2,2,1,5),(3,3,6,4)\}$ and let $B' = \{(3,1,1),(2,5,5),(1,2,3)\}$. Find the matrix of T with respect to B and B'.

2. Let B and B' be the bases from Exercise 1 above and let $S : R^4 \to R^3$ be the linear transformation whose matrix with respect to B and B' is $\begin{bmatrix} 2 & 1 & 1 & 4 \\ 3 & 2 & 1 & 1 \\ 5 & 3 & 2 & 5 \end{bmatrix}$. Find $S(x,y,z,w)$.

3. Find bases for the kernel and image of the linear transformations T from Exercise 1 and S from exercise 2.

4. **For students who have studied calculus.** Let $D : P_3 \to P_3$ be the linear transformation defined by $D(p) = \dfrac{dp}{dx}$ (the derivative of $p(x)$), and let $B = \{1,\ 1+x,\ x+x^2,\ x^2+x^3\}$. Find the matrix of D with respect to B.

5. Let $T : R^3 \to M_{2,2}$ be the linear transformation defined by $T(x,y,z) = \begin{bmatrix} x+y & x-z \\ y+z & x+y+z \end{bmatrix}$. Let $B = \{(1,3,3),(2,1,4),(3,2,5)\}$ and let $B' = \left\{ \begin{bmatrix} 2 & 1 \\ 3 & 1 \end{bmatrix}, \begin{bmatrix} 1 & 1 \\ 2 & 3 \end{bmatrix}, \begin{bmatrix} 3 & 2 \\ 1 & 3 \end{bmatrix}, \begin{bmatrix} 1 & 7 \\ 3 & 2 \end{bmatrix} \right\}$. Find the matrix of T with respect to B and B'.

6. Let $T : P_3 \to R^3$ be defined by $T(p) = (p(1)+p(2), p(1)-p(2), p(1))$. Find bases for the image and kernel of T.

7. Let $T : R^4 \to R^4$ be the linear transformation defined by $T(x,y,z,w) = (11x+4y+4z+12w, 18x+4y+8z+18w, 12x+2y+7z+12w, -18x-5y-8z-19w)$. Find a basis B such that the matrix of T with respect to B is diagonal.

8. Let $T : P_3 \to P_3$ be defined by $T(ax^3+bx^2+cx+d) = (8a+2b+2c+6d)x^3+(2z+3b+2d+2c)x^2+(-2a-2b+5c-2d)x+(-4a-b-3c-2d)$. Find a basis B for P_3 such that the matrix of T with respect to B is diagonal.

9. Determine if the matrices $\begin{bmatrix} 1 & 2 & 1 & 1 & 3 \\ 2 & 1 & 4 & 2 & 3 \\ 1 & 2 & 3 & 1 & 2 \\ 3 & 1 & 1 & 2 & 3 \\ 2 & 2 & 2 & 3 & 1 \end{bmatrix}$ and $\begin{bmatrix} 0 & 6 & -1 & 8 & 0 \\ 1 & 3 & 0 & 6 & 2 \\ 0 & 8 & -1 & 11 & -1 \\ -1 & 5 & 0 & 5 & -1 \\ 2 & 5 & -2 & 6 & 1 \end{bmatrix}$ are similar.

10. Determine if the matrices $\begin{bmatrix} 1 & 2 & 3 & 2 \\ 1 & 2 & 4 & 2 \\ 2 & 2 & 1 & 1 \\ 2 & 1 & 2 & 3 \end{bmatrix}$ and $\begin{bmatrix} -215 & -90 & -60 & -252 \\ 216 & 91 & 60 & 252 \\ -432 & -180 & -119 & -504 \\ 216 & 90 & 60 & 253 \end{bmatrix}$ are similar.

Exploration and Discovery

1. Let $A = \begin{bmatrix} 11 & 4 & 4 & 12 \\ 18 & 4 & 8 & 18 \\ 12 & 2 & 7 & 12 \\ -18 & -5 & -8 & -19 \end{bmatrix}$, let $P = \begin{bmatrix} 1 & 1 & 1 & 0 \\ 0 & 1 & 1 & 1 \\ 1 & 0 & 1 & 1 \\ 1 & 1 & 0 & 1 \end{bmatrix}$, and let $B = PAP^{-1}$. Then by definition, A is similar to B. Using A and B, test the following statements. Make up new matrices A and B of your own and test the statements a second time. Give counterexamples to those you determine to be false. Try to prove those that are true.

   (a) Similar matrices have the same determinant.

   (b) Similar matrices have the same reduced echelon form.

   (c) Similar matrices have the same characteristic polynomial.

   (d) Similar matrices have the same eigenvalues.

   (e) Similar matrices have the same eigenspaces.

LABORATORY EXERCISE 17.1

Matrices and Linear Transformations I

Name ______________________ **Due Date** ______________________

Let $T : P_2 \to P_2$ be a linear transformation such that

$T(x^2 - x) = 50x^2 - 21x + 65$, $T(x^2) = 5x^2 - 2x + 6$, and $T(x + 1) = -26x^2 + 11x - 34$

1. Find the matrix, $[T]$, of T with respect to the standard basis of P_2.

2. Find the matrix of T with respect to the basis $\{x^2 - x, x^2, x + 1\}$.

3. Determine if $[T]^{-1}$ exists.

4. Determine if T^{-1} exists. If it does, find $T^{-1}(ax^2 + bx + c)$.

LABORATORY EXERCISE 17.2

Matrices and Linear Transformations II

Name ______________________________ Due Date ________________

Let $T : P^3 \rightarrow P_3$ be a linear transformation such that

$$T(x^3 - x) = x^2 + 5,\ T(x^3 + 1) = x^2 - 1,\ T(x^2) = 3, \text{ and } T(x^2 - 1) = x^2 - x$$

1. Find the matrix of T with respect to the standard basis of P_3.

2. Find the matrix of T with respect to the basis $\{x^3 - x,\ x^3 + 1,\ x^2,\ x^2 - 1\}$.

3. Find a basis for the kernel of T.

4. Find a basis for the image of T.

CHAPTER 18

APPLICATIONS AND NUMERICAL METHODS

LINEAR ALGEBRA CONCEPTS

- **Systems of differential equations**
- **Gauss-Seidel method**
- **Generalized inverse and curve fitting**
- **Rotation of axes**
- ***LU* and *QR* factorizations**

Introduction

We began our study of linear algebra by solving systems of linear equations. In this chapter we look at how the ideas we have developed are used in solving systems of linear *differential* equations.

In some applications it may be necessary to solve huge systems of equations that are too large even for *DERIVE* to handle. (Although we should mention that there is a new version called *DERIVE–XM* that accesses *gigabytes* of extended memory!) In addition, the number of calculations required to row-reduce the augmented matrix for a very large system may require an impractical amount of time on even the fastest supercomputers and can lead to significant round-off error. For these reasons, other methods must be used to approximate solutions of large systems. We illustrate one such method known as *Gauss-Seidel iteration.*

Sometimes there may not be a solution to a system of equations, but there may be something very close to it. For example, there may not be a straight line passing through three given points, but there may be one that comes close. We explore this idea.

Certain operations on a matrix can result in a factorization of the matrix. We look at the two most common of these, the *LU* and *QR* factorizations.

Solved Problems

<u>Solved Problem 1</u>: **For students who have studied calculus.** Solve the following system of linear differential equations.

$$\begin{aligned}\frac{dx_1}{dt} &= -8x_1 + 4x_2 + 9x_3 \\ \frac{dx_2}{dt} &= 33x_1 - 9x_2 - 27x_3 \\ \frac{dx_3}{dt} &= -24x_1 + 6x_2 + 19x_3\end{aligned}$$

<u>Discussion</u>: $x_1 = x_1(t), x_2 = x_2(t)$, and $x_3 = x_3(t)$ are functions of t. Let $x' = \frac{dx}{dt}$, $\mathbf{x} = \begin{bmatrix} x_1 \\ x_2 \\ x_3 \end{bmatrix}$, and $A = \begin{bmatrix} -8 & 4 & 9 \\ 33 & -9 & -27 \\ -24 & 6 & 19 \end{bmatrix}$. We can then write the system of linear differential equations above as $\mathbf{x}' = A\mathbf{x}$.

If A happened to be a diagonal matrix, $\begin{bmatrix} p & 0 & 0 \\ 0 & q & 0 \\ 0 & 0 & r \end{bmatrix}$, the system of equations would be

$$\begin{aligned}x_1'(t) &= px_1(t) \\ x_2'(t) &= qx_2(t) \\ x_3'(t) &= rx_3(t)\end{aligned}$$

This system is called *uncoupled* because each equation involves a single variable that does not appear in the others. Therefore, each equation can be solved separately to yield

$$\begin{aligned}x_1(t) &= ae^{pt} \\ x_2(t) &= be^{qt} \\ x_3(t) &= ce^{rt}\end{aligned}$$

Since A is not a diagonal matrix, it is not obvious how to solve the given system; but suppose A is diagonalizable; that is, $P^{-1}AP = D$, where D is a diagonal matrix. The idea is to make a substitution to *uncouple* the system. Let $\mathbf{y} = \begin{bmatrix} y_1(t) \\ y_2(t) \\ y_3(t) \end{bmatrix}$ and put $\mathbf{x} = P\mathbf{y}$.

Then

$$\begin{aligned} \mathbf{x}' &= A\mathbf{x} \\ P\mathbf{y}' &= AP\mathbf{y} \\ \mathbf{y}' &= (P^{-1}AP)\mathbf{y} \\ \mathbf{y}' &= D\mathbf{y} \end{aligned}$$

This last system is uncoupled and we can solve it as above. Once $\mathbf{y}$ has been found, we can recover $\mathbf{x}$ using $\mathbf{x} = P\mathbf{y}$.

Solution of the problem: Using the methods of Chapter 15, we find P so that $P^{-1}AP = D$ is diagonal. We will omit the steps, but if $P = \begin{bmatrix} 1 & -1 & 3 \\ -3 & 4 & -9 \\ 2 & -3 & 7 \end{bmatrix}$ then $P^{-1}AP = \begin{bmatrix} -2 & 0 & 0 \\ 0 & 3 & 0 \\ 0 & 0 & 1 \end{bmatrix}$. If we solve the uncoupled system $\mathbf{y}' = D\mathbf{y}$ we get

$$\begin{aligned} y_1(t) &= ae^{-2t} \\ y_2(t) &= be^{3t} \\ y_3(t) &= ce^{t} \end{aligned}$$

Now, $\mathbf{x} = P\mathbf{y} = \begin{bmatrix} 1 & -1 & 3 \\ -3 & 4 & -9 \\ 2 & -3 & 7 \end{bmatrix} \begin{bmatrix} ae^{-2t} \\ be^{3t} \\ ce^{t} \end{bmatrix} = \begin{bmatrix} ae^{-2t} - be^{3t} + 3ce^{t} \\ -3ae^{-2t} + 4be^{3t} - 9ce^{t} \\ 2ae^{-2t} - 3be^{3t} + 7ce^{t} \end{bmatrix}$.

This process is seen in Figure 18.1. Note the order in which the functions appear in expression 3.

Once we have arrived at a final answer, we should check it as follows: To get the right-hand side of the original system, **Simplify** expression 4 in Figure 18.2, which results in expression 5. To get the left-hand side of the original system, highlight expression 3 in Figure 18.1 and use **Calculus Differentiate**. The result appears as expression 6 in Figure 18.2. **Simplify** expression 6 in Figure 18.2 to get expression 7. Expressions 5 and 7 are the same, although they go off the screen. (Use Ctrl → to scroll across.) We may also verify that they are the same by observing that we get 0 if we **Author** and **Simplify** #5-#7.

1: $$\begin{bmatrix} 1 & -1 & 3 \\ -3 & 4 & -9 \\ 2 & -3 & 7 \end{bmatrix}$$

2: $$\begin{bmatrix} 1 & -1 & 3 \\ -3 & 4 & -9 \\ 2 & -3 & 7 \end{bmatrix} \cdot \left[a\,ê^{-2t},\ b\,ê^{3t},\ c\,ê^{t}\right]$$

3: $$\left[-b\,ê^{3t} + 3\,c\,ê^{t} + a\,ê^{-2t},\ 4\,b\,ê^{3t} - 9\,c\,ê^{t} - 3\,a\,ê^{-2t},\ -3\,b\,ê^{3t}\right.$$

Figure 18.1: The solution of the system of differential equations

---

4: $$\begin{bmatrix} -8 & 4 & 9 \\ 33 & -9 & -27 \\ -24 & 6 & 19 \end{bmatrix} \cdot \left[-b\,ê^{3t} + 3\,c\,ê^{t} + a\,ê^{-2t},\ 4\,b\,ê^{3t} - 9\,c\,ê^{t} - 3\,a\right.$$

5: $$\left[-3\,b\,ê^{3t} + 3\,c\,ê^{t} - 2\,a\,ê^{-2t},\ 12\,b\,ê^{3t} - 9\,c\,ê^{t} + 6\,a\,ê^{-2t},\ -9\,b\right.$$

6: $$\frac{d}{dt}\left[-b\,ê^{3t} + 3\,c\,ê^{t} + a\,ê^{-2t},\ 4\,b\,ê^{3t} - 9\,c\,ê^{t} - 3\,a\,ê^{-2t},\ -3\,b\,ê^{3}\right.$$

7: $$\left[-3\,b\,ê^{3t} + 3\,c\,ê^{t} - 2\,a\,ê^{-2t},\ 12\,b\,ê^{3t} - 9\,c\,ê^{t} + 6\,a\,ê^{-2t},\ -9\,b\right.$$

Figure 18.2: Checking the solution of the system

Solved Problem 2: **This problem requires the LU.MTH file in Appendix II.** Use the *LU* decomposition to solve the following system of equations.

$$\begin{aligned} 2x_1 + 6x_2 + 4x_3 + 2x_4 &= 2 \\ x_1 + 6x_2 + 11x_3 + 7x_4 &= 4 \\ 3x_1 + 11x_2 + 17x_3 + 12x_4 &= 10 \\ 2x_1 + 10x_2 + 18x_3 + 13x_4 &= 9 \end{aligned}$$

Discussion: Scientists and engineers often encounter systems of equations that are large enough to make computation time and round-off error significant issues. A number of methods have been developed to deal with this problem. We will look at two: *LU* (Lower-Upper) factorization in Solved Problem 2 and Gauss-Seidel iteration in Solved Problem 3.

***LU* Factorization Theorem:** If A is a matrix that can be put into echelon form without row interchanges, then there is a lower triangular matrix L and an upper triangular matrix U such that $A = LU$.

The code in the LU.MTH section of Appendix II will produce the required decomposition. We will use it to illustrate how the *LU* decomposition is used to solve a system of equations.

Solution: *DERIVE* can solve this system directly, but we are asked to use the *LU* decomposition. The first step is to write the system in matrix form.

$$\begin{bmatrix} 2 & 6 & 4 & 2 \\ 1 & 6 & 11 & 7 \\ 3 & 11 & 17 & 12 \\ 2 & 10 & 18 & 13 \end{bmatrix} \begin{bmatrix} x_1 \\ x_2 \\ x_3 \\ x_4 \end{bmatrix} = \begin{bmatrix} 2 \\ 4 \\ 10 \\ 9 \end{bmatrix}$$

Enter the coefficient matrix using **Declare Matrix** and name it A. The next step is to obtain the *LU* factorization of A. Use **Transfer Load Utility** to call the LU.MTH file; then **Author** and **Simplify** the expressions `lower(A)` and `upper(A)`. The results are seen as expressions 27 and 29 of Figure 18.3. (The reader should use *DERIVE* to verify that the product LOWER(A) UPPER(A) does in fact equal A.)

With this factorization we may write our system of equations as follows.

$$\begin{bmatrix} 2 & 0 & 0 & 0 \\ 1 & 3 & 0 & 0 \\ 3 & 2 & 5 & 0 \\ 2 & 4 & 2 & 1 \end{bmatrix} \left(\begin{bmatrix} 1 & 3 & 2 & 1 \\ 0 & 1 & 3 & 2 \\ 0 & 0 & 1 & 1 \\ 0 & 0 & 0 & 1 \end{bmatrix} \begin{bmatrix} x_1 \\ x_2 \\ x_3 \\ x_4 \end{bmatrix} \right) = \begin{bmatrix} 2 \\ 4 \\ 10 \\ 9 \end{bmatrix}$$

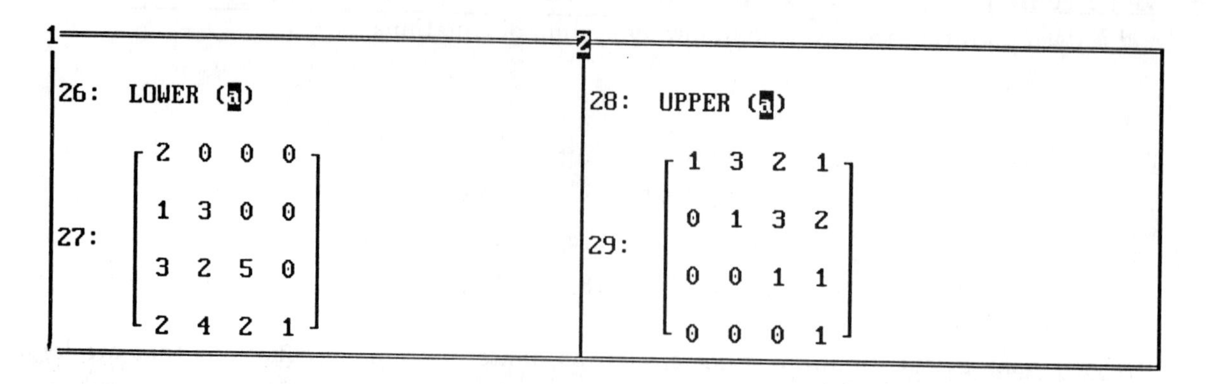

Figure 18.3: **The *LU* decomposition of a matrix**

Introducing new variables, we can replace this system of equations by two simpler systems of equations.

$$\begin{bmatrix} 2 & 0 & 0 & 0 \\ 1 & 3 & 0 & 0 \\ 3 & 2 & 5 & 0 \\ 2 & 4 & 2 & 1 \end{bmatrix} \begin{bmatrix} y_1 \\ y_2 \\ y_3 \\ y_4 \end{bmatrix} = \begin{bmatrix} 2 \\ 4 \\ 10 \\ 9 \end{bmatrix} \tag{18.1}$$

$$\begin{bmatrix} 1 & 3 & 2 & 1 \\ 0 & 1 & 3 & 2 \\ 0 & 0 & 1 & 1 \\ 0 & 0 & 0 & 1 \end{bmatrix} \begin{bmatrix} x_1 \\ x_2 \\ x_3 \\ x_4 \end{bmatrix} = \begin{bmatrix} y_1 \\ y_2 \\ y_3 \\ y_4 \end{bmatrix} \tag{18.2}$$

Rewrite equation (18.1) as a system of equations:

$$\begin{aligned} 2y_1 &= 2 \\ y_1 + 3y_2 &= 4 \\ 3y_1 + 2y_2 + 5y_3 &= 10 \\ 2y_1 + 4y_2 + 2y_3 + y_4 &= 9 \end{aligned}$$

This system is easily solved by *forward substitution* to yield $y_1 = 1$, $y_2 = 1$, $y_3 = 1$, $y_4 = 1$. Using this solution we write equation (18.2) as a system of equations.

$$\begin{aligned} x_1 + 3x_2 + 2x_3 + x_4 &= 1 \\ x_2 + 3x_3 + 2x_4 &= 1 \\ x_3 + x_4 &= 1 \\ x_4 &= 1 \end{aligned}$$

We solve this system by *back substitution* to yield, $x_4 = 1$, $x_3 = 0$, $x_2 = -1$, $x_1 = 4$. (The reader should check the answers with *DERIVE.*)

For a small system of equations, this is not an efficient method, but for a large system, the *LU* factorization allows a complicated system of equations to be replaced by two systems of equations that can be easily solved by forward and backward substitution. Very efficient computer code exists to implement these procedures.

Solved Problem 3: (Gauss-Seidel iteration.) Let $A = \begin{bmatrix} 5 & 2 & 1 & 1 \\ 3 & 6 & 3 & 1 \\ 2 & 2 & 7 & 3 \\ 1 & 2 & 3 & 6 \end{bmatrix}$ and $\mathbf{b} = (4, 3, 2, 1)$.

Use ten iterations of the Gauss-Seidel method to approximate the solution of the system of equations $A \cdot \mathbf{x} = \mathbf{b}$.

Discussion: In some cases it is better to approximate the solution of a system of equations rather than attempt to solve it exactly. The Gauss-Seidel method is such an approximation technique. It is a refinement of a similar one called the Jacobi method, about which we'll say a bit more later.

For this system of equations the exact solution in expression 5 of Figure 18.4 is easily obtained using *DERIVE.* The purpose of this problem is only to illustrate how the Gauss-Seidel method works. What we will present is a simplified version of the actual procedure used in practice.

Solution: Let M denote the lower triangular part of A; that is, $M = \begin{bmatrix} 5 & 0 & 0 & 0 \\ 3 & 6 & 0 & 0 \\ 2 & 2 & 7 & 0 \\ 1 & 2 & 3 & 6 \end{bmatrix}$.

Then the original system of equations can be written as $M \cdot \mathbf{x} = (M - A) \cdot \mathbf{x} + \mathbf{b}$, or $\mathbf{x} = (I - M^{-1}A) \cdot \mathbf{x} + M^{-1} \cdot \mathbf{b}$. The idea is to begin with a guess at the solution (even a bad one such as $\mathbf{x_0} = (0,0,0,0)$) and improve it using the formula

$$\mathbf{x_{n+1}} = (I - M^{-1}A) \cdot \mathbf{x_n} + M^{-1} \cdot \mathbf{b}.$$

```
1:   a :=  [ 5  2  1  1 ]
           [ 3  6  3  1 ]
           [ 2  2  7  3 ]
           [ 1  2  3  6 ]

2:   b := [4, 3, 2, 1]

3:   a · [x, y, z, w] = b

4:   [5 x + 2 y + z + w = 4, 3 x + 6 y + 3 z + w = 3, 2 x + 2 y + 7 z + 3 w =

          131        71        17           1
5:   [x = ---, y = ---, z = ---, w = - --]
          174        696       348          58

6:   [x = 0.752873, y = 0.102011, z = 0.0488505, w = -0.0172413]
```

Figure 18.4: **Preparing for Gauss-Seidel iteration**

With M, A, and $\mathbf{b}$ defined as above and `I:=identity_matrix(4)`, **Author**

```
iterates((I-M^(-1).a).x+M^(-1).b,x,[0,0,0,0],10)
```

as seen in Figure 18.5. This tells *DERIVE* to begin with $\mathbf{x} = (0,0,0,0)$ and replace it with $(I-M^{-1}A)\cdot\mathbf{x}+M^{-1}\cdot\mathbf{b}$. This new value of $\mathbf{x}$ is plugged back into the formula to get a third value of $\mathbf{x}$. The procedure is repeated 10 times. We **approX** expression 10 of Figure 18.5 to get the matrix in expression 11. The rows of this matrix are successive approximations to a solution of the system. Compare the last row of the matrix in expression 11 of Figure 18.5 with the approximate solution of the system in expression 6 of Figure 18.4.

Remark on the Jacobi method: It is easy to see that our derivation of the formula $\mathbf{x} = (I - M^{-1}A)\cdot\mathbf{x} + M^{-1}\cdot\mathbf{b}$ from $A\mathbf{x} = \mathbf{b}$ holds if M is *any* invertible matrix! The Jacobi method is the same procedure we just executed with M replaced by the diagonal of A: $\begin{bmatrix} 5 & 0 & 0 & 0 \\ 0 & 6 & 0 & 0 \\ 0 & 0 & 7 & 0 \\ 0 & 0 & 0 & 6 \end{bmatrix}$. It is a bit less efficient than the Gauss-Seidel method.

```
                    -1            -1
10:  ITERATES ((i - m   · a) · x + m   · b, x, [0, 0, 0, 0], 10)
```

```
     ┌ 0          0          0          0          ┐
     │ 0.8        0.1        0.0285714  -0.0142857 │
     │ 0.757142   0.109523   0.0442176  -0.0181405 │
     │ 0.750975   0.105427   0.0488025  -0.0180394 │
     │ 0.751676   0.102767   0.0493187  -0.0175278 │
11:  │ 0.752534   0.101994   0.0490749  -0.0172914 │
     │ 0.752845   0.101921   0.0489057  -0.0172343 │
     │ 0.752897   0.101971   0.0488524  -0.0172327 │
     │ 0.752887   0.102002   0.0488455  -0.0172380 │
     │ 0.752877   0.102011   0.048848   -0.0172407 │
     └ 0.752873   0.102012   0.0488499  -0.0172413 ┘
```

Figure 18.5: **Ten iterations of the Gauss-Seidel method**

Solved Problem 4: (The generalized inverse and curve fitting.) In an experiment the following data were collected:

x	1	1.5	2	2.5	3
y	3.4	3.8	3.9	4.3	4.5

(a) Find the least squares straight line fit to this data and plot both the line and the data.

(b) Find the best least squares fit to this data by a function of the form $a \sin x + b \cos x + cx$. Calculate the (approximate) generalized inverse of the associated matrix and plot both the function and the data.

Discussion: If M is a matrix (not necessarily square) whose columns are linearly independent then it can be shown that $M^t M$ is invertible. (See Exploration and Discovery problem 2). When this is the case, $(M^t M)^{-1} M^t$ is called the *generalized (or pseudo-) inverse* of M.

If we want to solve a system of equations $M\mathbf{x} = \mathbf{y}$ and there are more equations than unknowns, there may not be a solution, so we look for the nearest thing to one - literally. To say the system $M\mathbf{x} = \mathbf{y}$ has no solution means that the vector $\mathbf{y}$ does not belong to the set of all vectors of the form $M\mathbf{x}$ (the column space of M); however, there *is* a unique vector in the column space that is nearest to it. That vector is $(M^tM)^{-1}M^t\mathbf{y}$.

Solution to (a): Ideally, we want to find a and b for which the line $a + bx$ goes through the data. That is,

$$\begin{aligned} a + b(1) &= 3.4 \\ a + b(1.5) &= 3.8 \\ a + b(2) &= 3.9 \\ a + b(2.5) &= 4.3 \\ a + b(3) &= 4.5 \end{aligned}$$

This means we want to solve $M \begin{bmatrix} a \\ b \end{bmatrix} = \mathbf{y}$, where $M = \begin{bmatrix} 1 & 1 \\ 1 & 1.5 \\ 1 & 2 \\ 1 & 2.5 \\ 1 & 3 \end{bmatrix}$ and $\mathbf{y} = \begin{bmatrix} 3.4 \\ 3.8 \\ 3.9 \\ 4.3 \\ 4.5 \end{bmatrix}$, but there is no solution. (Why?) Thus we find the best approximation with the generalized inverse $(M^tM)^{-1}M^t\mathbf{y}$. The line we seek then is $a + bx$ where $\begin{bmatrix} a \\ b \end{bmatrix} = (M^tM)^{-1}M^t\mathbf{y}$. We suggest using **Declare vectoR** to enter each of the vectors

$$[1, 1, 1, 1, 1], \ [1, 1.5, 2, 2.5, 3], \text{ and } [3.4, 3.8, 3.9, 4.3, 4.5]$$

separately as expressions 1, 2, and 3 in Figure 18.6.

DERIVE Remark: Entering matrices of data can be a chore. If your data have been generated by another program or if you have entered the data in a file, you can load it as a matrix with the **Transfer Load daTa** command. See Section 2.9 of your *DERIVE* manual. This feature is not implemented for versions earlier than 2.5.

Now **Author** `M:=[#1,#2]'` to define the matrix M. Note *DERIVE* 's transpose symbol ('); it will be used often, so be careful not to overlook it.

Next **Author** `(M'. M)^(-1). M'. #3`. When we **Simplify** it we get the vector $\mathbf{x}$ in expression 6 of Figure 18.6. A quick way to get the desired formula is to **Author** `#6.[1,x]` and **Simplify** it. The result is expression 8 in Figure 18.6.

Now we must plot it along with the data points. (See Section 13 in Appendix I for details on plotting.) **Author** `[#2,#3]'`, **Simplify** (not shown), and **Plot** to get the data points.

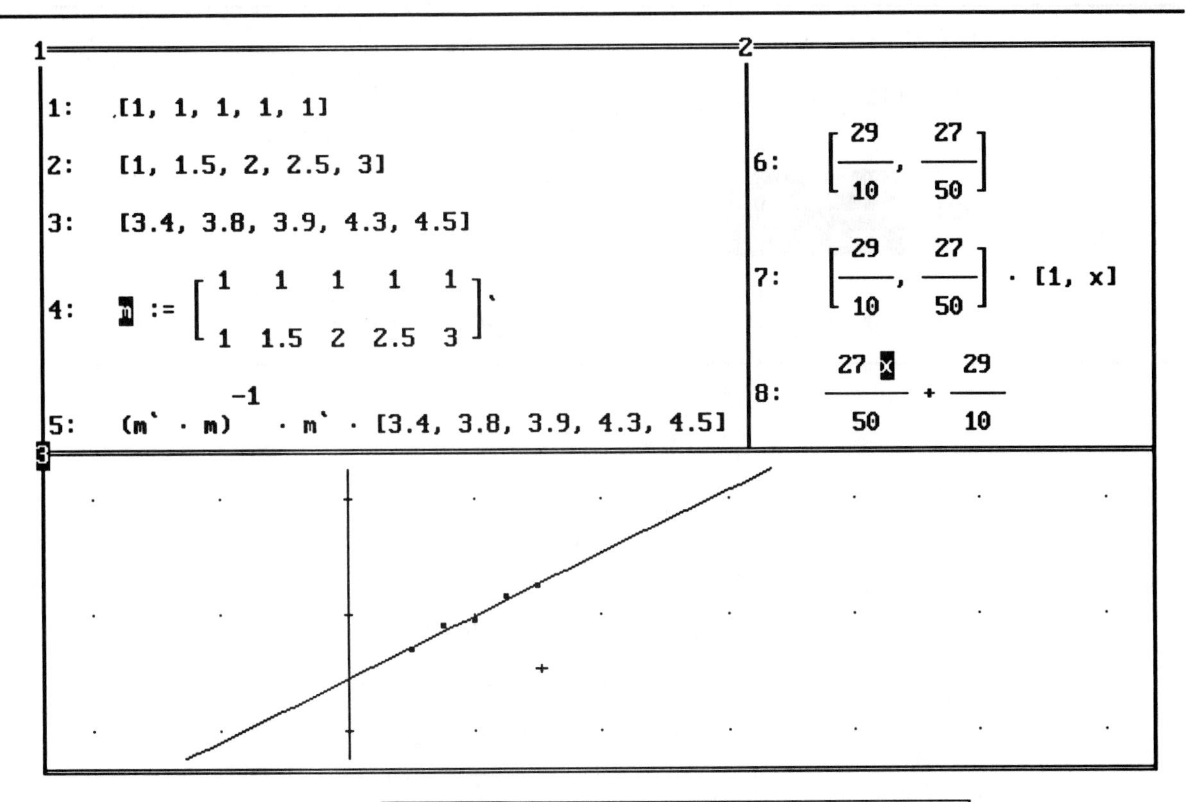

Figure 18.6: **Fitting a line to data. (Graph scale 2:2)**

(We moved the graphing cursor up and to the right and used the **Center** command to get the graph in Figure 18.6.) Return to the algebra screen with F1, highlight expression 8 and **Plot** again to get the line.

DERIVE's FIT function: *DERIVE* automates the process we just went through with the built-in function FIT. You can see the syntax in Figure 18.7. The matrix of data may be entered with the x-coordinates in column 1 and the y-coordinates in column 2, or in rows and then the transpose taken, as in expression 11 of Figure 18.7. The form of the function you want to fit to the data is entered in vector form. If you want to fit a general quadratic to the data, you would enter `[x, ax^2 + bx + c]`. You may fit any linear combination of functions.

Solution to (b): Ideally, we want to find a and b and c for which $a \sin x + b \cos x + cx$ goes through the data. That is,

```
11:  FIT [[x, a x + b], [ 1    1.5   2    2.5   3  ]']
                        [ 3.4  3.8   3.9  4.3   4.5]

      27 x     29
12:   ----  +  --
       50      10
```

Figure 18.7: **The FIT function for a line**

$$\begin{aligned} a\sin 1 + b\cos 1 + c &= 3.4 \\ a\sin 1.5 + b\cos 1.5 + 1.5c &= 3.8 \\ a\sin 2 + b\cos 2 + 2c &= 3.9 \\ a\sin 2.5 + b\cos 2.5 + 2.5c &= 4.3 \\ a\sin 3 + b\cos 3 + 3c &= 4.5 \end{aligned}$$

However, there is no solution to this system, so we must find the generalized inverse of the coefficient matrix to get a "best fit." We need to generate the vector

$$[\sin 1, \sin 1.5, \sin 2, \sin 2.5, \sin 3]$$

This is done most efficiently if we **Author** and **approX**

```
vector(sin(element(#1, i)),i,1,5)
```

as seen in expressions 4 and 5 of Figure 18.8. Do the same for the cosines. We **Author** `M:=[#1,#5,#7]'` to define the matrix M.

Next **Author** `(M'. M)^(-1). M'` and **approX**. The generalized inverse appears as expression 10 in Figure 18.9. Next **Author** `#10.#2` and **approX** to get expressions 11 and 12. The vector in expression 12 gives the coefficients of x, $\sin x$, and $\cos x$ in that order.

In expressions 13 and 14 we have approximated the FIT function to illustrate that its result agrees with ours. The plots of the data and the function appear in Figure 18.10. Window 4 shows a close-up of the graph in window 3.

1: [1, 1.5, 2, 2.5, 3]

2: [3.4, 3.8, 3.9, 4.3, 4.5]

3: $\begin{bmatrix} 1 & 1.5 & 2 & 2.5 & 3 \\ 3.4 & 3.8 & 3.9 & 4.3 & 4.5 \end{bmatrix}^{`}$

4: VECTOR (SIN (ELEMENT ([1, 1.5, 2, 2.5, 3], i)), i, 1, 5)

5: [0.84147, 0.997494, 0.909297, 0.598472, 0.14112]

6: VECTOR (COS (ELEMENT ([1, 1.5, 2, 2.5, 3], i)), i, 1, 5)

7: [0.540302, 0.0707373, -0.416146, -0.801143, -0.989992]

8: $m := \begin{bmatrix} 1 & 1.5 & 2 & 2.5 & 3 \\ 0.84147 & 0.997494 & 0.909297 & 0.598472 & 0.14112 \\ 0.540302 & 0.0707373 & -0.416146 & -0.801143 & -0.989992 \end{bmatrix}^{`}$

Figure 18.8: **Fitting $a \sin x + b \cos x + cx$ to data**

```
9:   (m` · m)^-1 · m`

       ┌ 0.550052  -0.0874759  -0.389547  -0.177526   0.601358 ┐
10:    │ -0.620375   0.515138    0.973979   0.460292  -1.08367  │
       └ 1.47456   -0.06558    -0.939589  -0.716784   0.764981 ┘

       ┌ 0.550052  -0.0874759  -0.389547  -0.177526   0.601358 ┐
11:    │ -0.620375   0.515138    0.973979   0.460292  -1.08367  │ · [3.4, 3.8, 3.9,
       └ 1.47456   -0.06558    -0.939589  -0.716784   0.764981 ┘

12:  [1.96128, 0.749508, 1.46014]

13:  FIT [[x, a SIN (x) + b COS (x) + c x], [ 1    1.5  2    2.5  3   ]` ]
                                              [ 3.4  3.8  3.9  4.3  4.5 ]

14:  1.46016 COS (x) + 0.749494 SIN (x) + 1.96127 x
```

Figure 18.9: **The generalized inverse and FIT for $a \sin x + b \cos x + cx$**

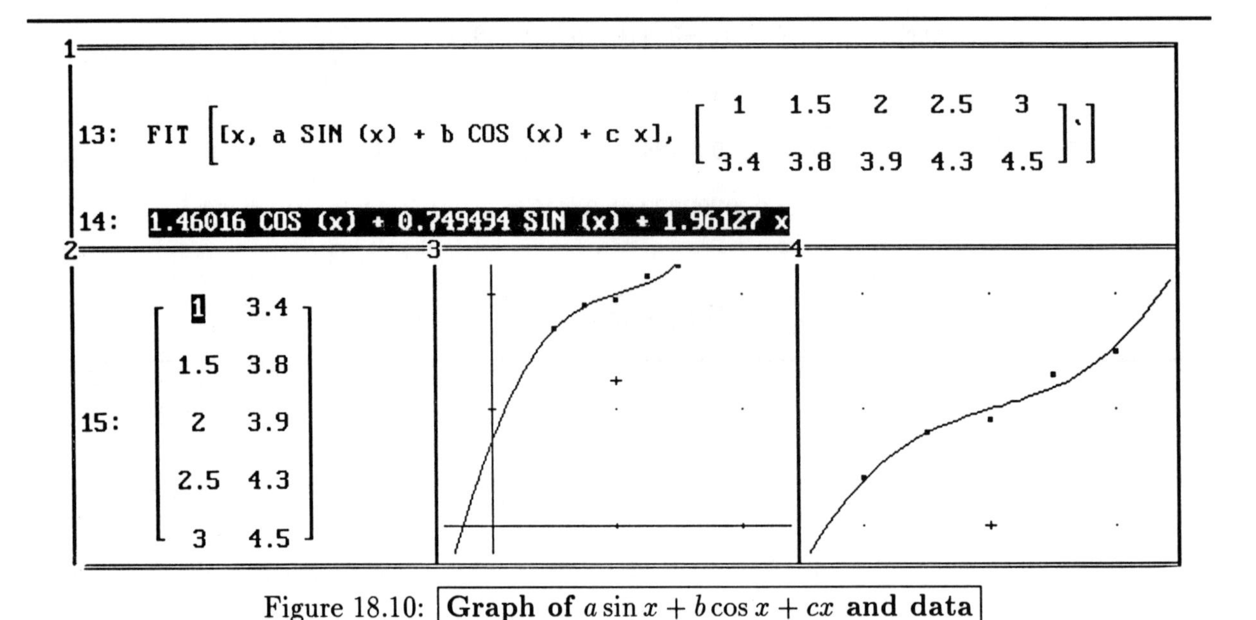

Figure 18.10: **Graph of $a \sin x + b \cos x + cx$ and data**

Solved Problem 5: Plot the graph of $4x^2 + 3xy + y^2 + 2x + y - 1$. Apply a rotation through 22.5 degrees and plot the graph again.

Solution: **Author** $4x^2 + 3xy + y^2 + 2x + y - 1$. *DERIVE* will not do implicit plots, so we must first **soLve** $4x^2 + 3xy + y^2 + 2x + y - 1$ for y, as in expressions 2 and 3 of Figure 18.11. The solutions are

$$y = -\frac{\sqrt{-7x^2 - 2x + 5} + 3x + 1}{2} \quad \text{and} \quad y = \frac{\sqrt{-7x^2 - 2x + 5} - 3x - 1}{2}$$

Highlight expression 2 and use **Plot Beside** to plot it. Press F1 to get back to the algebra window, highlight expression 3, and **Plot** it.

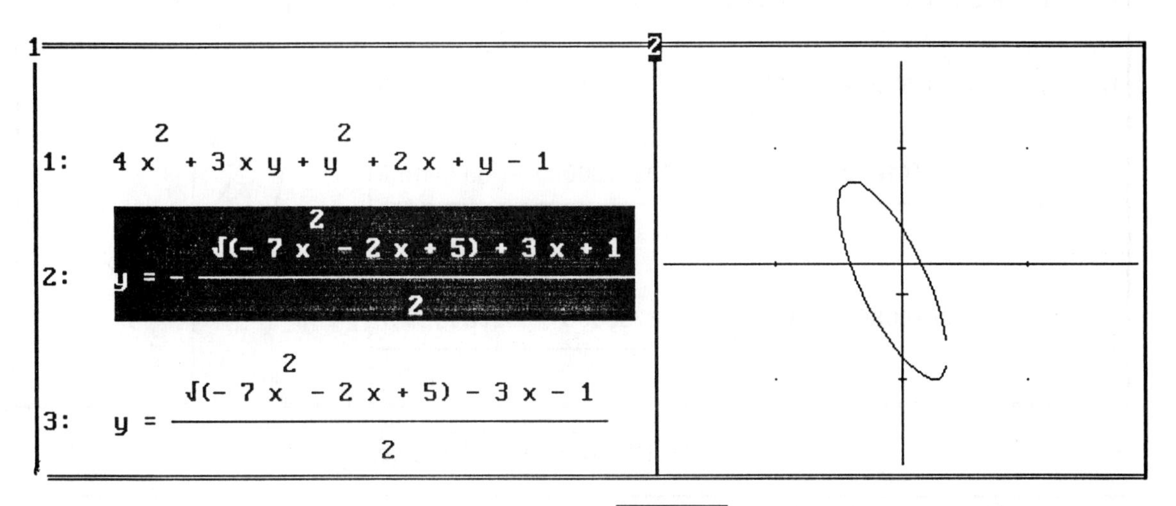

Figure 18.11: **Ellipse**

We see that we get two halves of an ellipse and there's a small gap where they come together. This is just a peculiarity of computer graphing. The gap should be smoothly filled in.

Next we use **Declare Matrix** to enter the rotation matrix $\begin{bmatrix} \cos(22.5^o) & -\sin(22.5^o) \\ \sin(22.5^o) & \cos(22.5^o) \end{bmatrix}$ as in expression 9 of Figure 18.12. (To enter 22.5 degrees, type `22.5 deg`.)

To get the rotation transformation, **Author** `#9.[x,y]` and **approX**. We obtain the result $[0.923879x - 0.382683y,\ 0.382683x + 0.923879y]$ as expression 11. Now we want to substitute the first coordinate of expression 11 for x in $4x^2 + 3xy + y^2 + 2x + y - 1$ and the second coordinate for y. Highlight expression 1 and use **Manage Substitute** to do this. (The left and right arrow keys will let us highlight the individual coordinates of expression 11.)

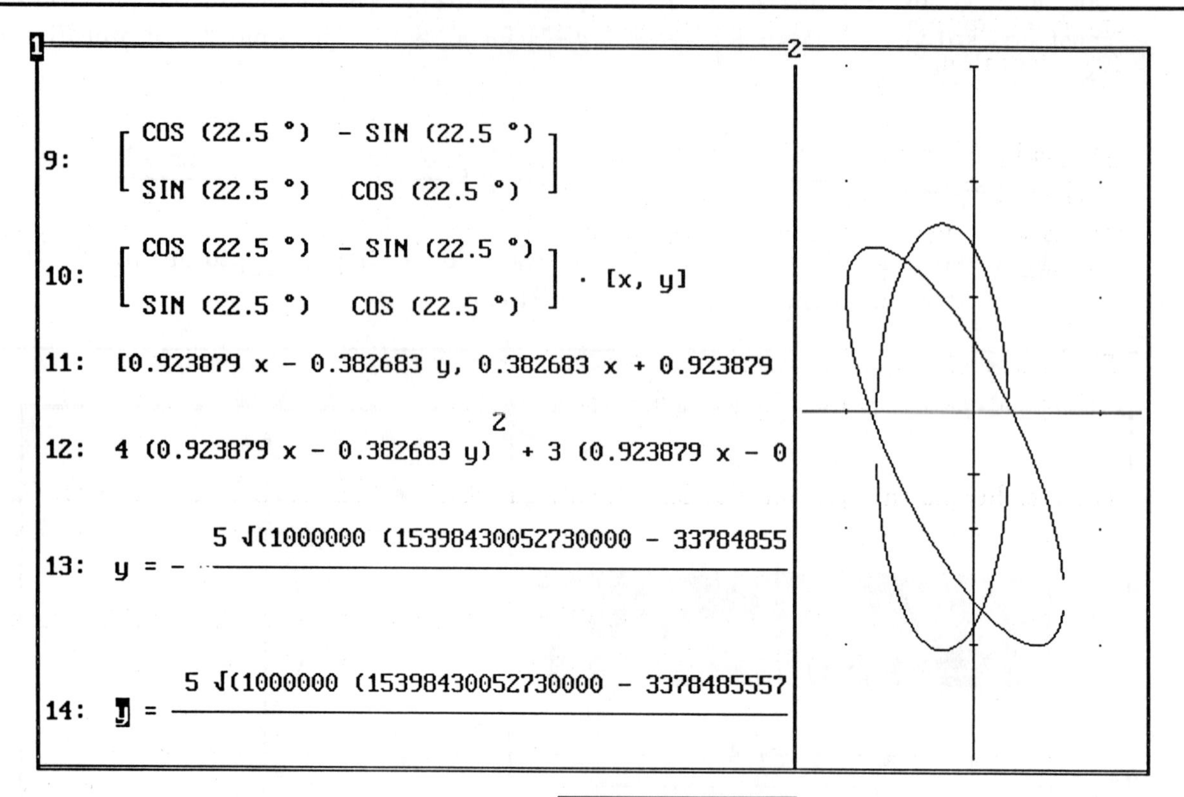

Figure 18.12: **Rotated Ellipse**

The result of the substitution is expression 12: $4(0.923879x - 0.382683y)^2 + 3(0.923879x - 0.382683y)(0.382683x + 0.923879y) + 1(0.382683x + 0.923879y)^2 + 2(0.923879x - 0.382683y) + 1(0.382683x + 0.923879y) - 1$, which is partially obscured by the plots.

We **soLve** this and we get two solutions, which we plot as before. This time we see the two halves of the rotated ellipse, again with gaps that should be filled in.

Solved Problem 6: **This exercise uses the file QR.MTH in Appendix II.**

Let $A = \begin{bmatrix} 1 & 3 & 5 \\ 2 & 3 & 1 \\ 2 & 4 & -1 \end{bmatrix}$. Find the QR factorization of A. Calculate the eigenvalues of A and compare them with approximations obtained by executing 5, 10, and 30 steps of the QR algorithm.

Discussion: It is not practical to find eigenvalues of large matrices by finding the roots of its characteristic polynomial. An algorithm known as the *QR algorithm* efficiently approximates the eigenvalues in many cases. The algorithm begins with the QR factorization theorem.

Theorem: **(QR factorization)** Let A be a nonsingular square matrix. Then there is an orthogonal matrix Q and a nonsingular upper triangular matrix R such that $A = QR$.

The idea behind the QR factorization is really quite simple. The columns of the matrix Q are just the result of applying the Gram-Schmidt algorithm to the columns of A and the entries of the matrix R contain the information necessary to keep track of the operations required to do this. There are many applications of the QR factorization, but among the most important is the following procedure known as the *QR algorithm.* We remark that this is a simplified version of what is used in practice.

The QR Algorithm

If A is a square nonsingular matrix, obtain the QR factorization, $A = QR$, and let $A_1 = RQ$ (the product of the QR factorization in reverse order).

Continue the process: A_2 is the QR factorization of A_1 taken in reverse order, and so on. In general, A_n is the result of applying this procedure n times.

For a matrix with real eigenvalues some fairly general conditions will ensure that the QR algorithm converges rapidly to a matrix that is *approximately upper triangular,* (i.e., the entries below the main diagonal have very small magnitude) and the entries on the main diagonal are approximations of the eigenvalues of A.

In the QR.MTH section of Appendix II we have provided *DERIVE* code that calculates the QR factorization and executes the QR algorithm. We use this code in the following solution.

Solution: First **Transfer Load Utility** the file QR.MTH. Enter the matrix and name it A. **Author** and **Simplify** the expressions `Q(A)` and `R(A)`. The results are seen in expressions 16 and 18 of Figure 18.13.

The matrix $R(A)$ is obviously upper triangular. The reader should use *DERIVE* to verify at least 1 and 3 below:

1. $Q(A)$ is an orthogonal matrix.
2. The columns of $Q(A)$ are the result of applying the Gram-Schmidt algorithm to the columns of A.
3. $Q(A)R(A) = A$.

Window 1:

15: $Q\begin{bmatrix} 1 & 3 & 5 \\ 2 & 3 & 1 \\ 2 & 4 & -1 \end{bmatrix}$

16: $\begin{bmatrix} \frac{1}{3} & \frac{10\sqrt{17}}{51} & \frac{2\sqrt{17}}{17} \\ \frac{2}{3} & -\frac{7\sqrt{17}}{51} & \frac{2\sqrt{17}}{17} \\ \frac{2}{3} & \frac{2\sqrt{17}}{51} & -\frac{3\sqrt{17}}{17} \end{bmatrix}$

Window 2:

17: $R\begin{bmatrix} 1 & 3 & 5 \\ 2 & 3 & 1 \\ 2 & 4 & -1 \end{bmatrix}$

18: $\begin{bmatrix} 3 & \frac{17}{3} & \frac{5}{3} \\ 0 & \frac{\sqrt{17}}{3} & \frac{41\sqrt{17}}{51} \\ 0 & 0 & \frac{15\sqrt{17}}{17} \end{bmatrix}$

Figure 18.13: **The QR factorization of A**

If we **Author** `charpoly(A)`, **soLve** , and **approX**, we find the eigenvalues of A are -0.845030, 6.55359, and -2.70856.

Even with *DERIVE* to help, it is tedious to execute the individual steps of the *QR algorithm*, but the function $\text{QR}(A, n)$ provided in the QR.MTH file automates the procedure. **Author** and **approX** the expression `qr(A,5)`. (*Warning*): **approX** , do not **Simplify**.) The result appears in window 2 of Figure 18.14. Observe that the entries below the main diagonal are relatively small in magnitude and that the entries on the main diagonal are close to the eigenvalues we calculated earlier. Notice also in windows 4 and 6 of Figure 18.14 that the approximations become much better as the number of iterations increases. A more refined version of this algorithm will converge more rapidly.

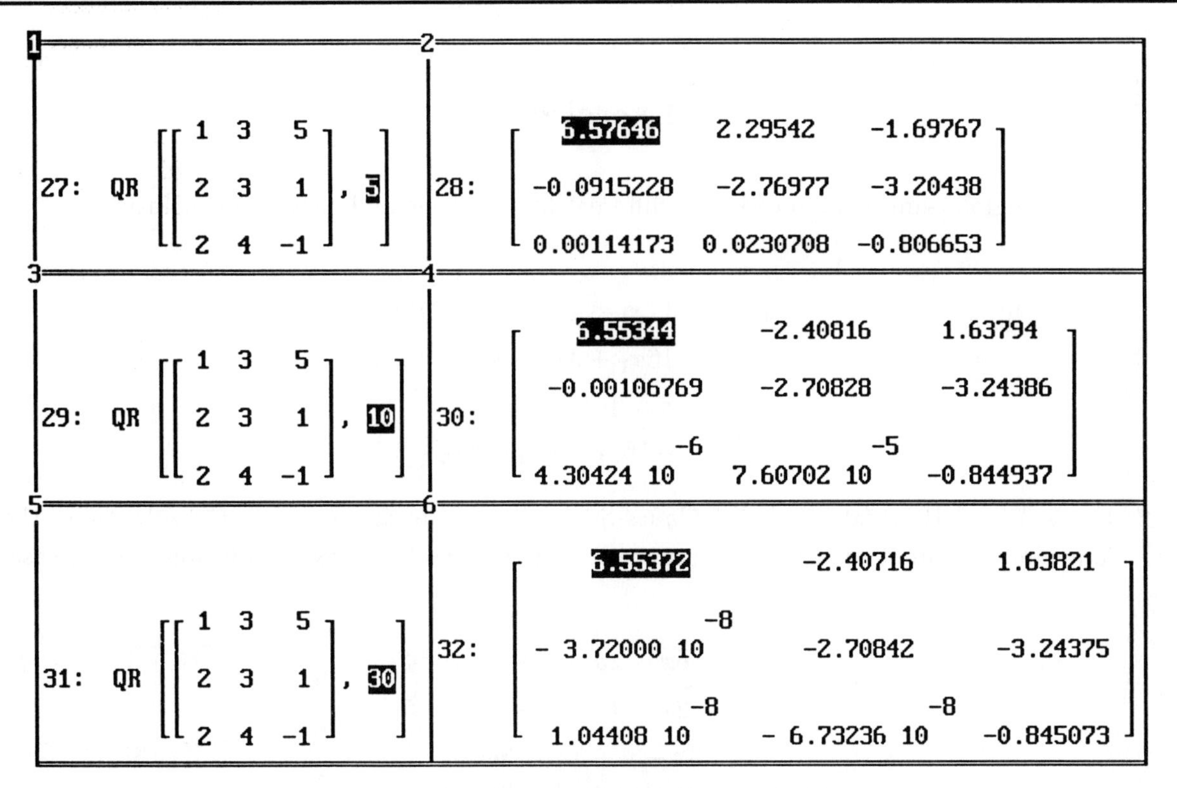

Figure 18.14: **The QR algorithm applied to A**

Exercises

1. Solve the following system of linear differential equations and check your solution.

$$\begin{aligned} x_1' &= 11x_1 + 4x_2 + 4x_3 + 12x_4 \\ x_2' &= 16x_1 + 3x_2 + 8x_3 + 16x_4 \\ x_3' &= 8x_1 + 7x_3 + 8x_4 \\ x_4' &= -16x_1 - 4x_2 - 8x_3 - 17x_4 \end{aligned}$$

2. Use 3, 7, and 10 iterations of the Gauss-Seidel method to approximate the solution of the following system of equations. Compare your answer with the exact solution in each case.

$$\begin{aligned} 6x + 2y - z + w &= 4 \\ 2x + 7y + z - w &= 5 \\ 3x - y + 5z + 2w &= 1 \\ 4x + 3y + 2z - 8w &= 2 \end{aligned}$$

3. Use the LU factorization to solve the system of equations in Exercise 2. Show all the steps in your solution.

4. Show that if M has an inverse, then $(M^tM)^{-1}M^t = M^{-1}$ (no computer for this).

5. Find the best fit to the data below by a function of the form $ax + b\sin x + c\cos x$. Show the matrix M and its generalized inverse. Plot the function and data points.

x	3	3.5	4	4.5	5	5.5	6	6.5	7	7.5	8
y	1.3	1.4	1.6	1.5	1.5	1.4	1.5	1.5	1.4	1.2	1.3

6. Find the best fit to the data above by a function of the form $ax + b\ln x$. Show the matrix M and its generalized inverse. Plot the function and data points.

7. Repeat Solved Problem 5 for $4x^2 - xy + y^2 + 2x + y - 1 = 0$.

8. Repeat Solved Problem 5 for $4x^2 - xy + y^2 + 2x + y - 1 = 0$ using an angle of $60°$.

Exploration and Discovery

1. Implement the Jacobi method discussed in Solved Problem 3 and apply it to Exercise 2. Compare the Jacobi and Gauss-Seidel methods step by step for several iterations in this problem and discuss your observations. Compare the methods for a 3 by 3 system.

2. This is not a computer problem. It's all pencil and paper. Suppose M is m by n and has linearly independent columns. Prove that M^tM has an inverse by the following steps:

   (a) Show that M^tM is square and therefore it suffices to show that if $(M^tM)\mathbf{x} = \mathbf{0}$, then $\mathbf{x} = 0$.

   (b) Show that if $\mathbf{u}$ is in R^m and $\mathbf{v}$ is in R^n then $(M^t\mathbf{u})\cdot\mathbf{v} = \mathbf{u}\cdot(M\mathbf{v})$.

   (c) Let $\mathbf{x}$ be in R^n and substitute **u**=M**x** and **v**=**x** in (b) to conclude that $|M\mathbf{x}|^2 = 0$.

   (d) Use (c) and the fact that M has linearly independent columns to conclude that $\mathbf{x} = 0$ as desired.

3. (a) Do Exercise 1.

   (b) Do Exploration and Discovery Problem 3(d) in Chapter 3.

   (c) Do Exploration and Discovery Problem 1(d) in Chapter 15 for e^x.

   (d) Let $\mathbf{c}$ be the vector of constants (c_1, c_2, c_3, c_4). Show that the solution of the system in Exercise 1 can be written as $\mathbf{x} = e^{At}\mathbf{c}$, where A is the coefficient matrix.

4. Rotating $ax^2 + bxy + cy^2 + dx + ey + k = 0$ through the proper angle will eliminate the xy term. Let's find such an angle.

   (a) Enter the rotation matrix $\begin{bmatrix} \cos\theta & -\sin\theta \\ \sin\theta & \cos\theta \end{bmatrix}$ and the quadratic $ax^2 + bxy + cy^2 + dx + ey + k$. (Use Alt H for θ.)

   (b) The linear transformation $T(x, y) = \begin{bmatrix} \cos\theta & -\sin\theta \\ \sin\theta & \cos\theta \end{bmatrix}\begin{bmatrix} x \\ y \end{bmatrix}$ rotates the vector (x, y) through the angle θ. Use **Manage Substitute** to substitute the first coordinate of $T(x, y)$ into $ax^2 + bxy + cy^2 + dx + ey + k$ for x and the second for y.

(c) After you **Simplify** the preceding substitution, use the left and right arrow keys to extract all the terms with a factor of xy. (Be sure you get them all.)

(d) Now, set the coefficient of xy equal to zero and **soLve** for θ. Discuss your observations and conclusions.

(e) Plot $4x^2 - xy + y^2 + 2x + y - 1 = 0$ as we did in Solved Problem 5.

(f) Use your result in part (d) to find an angle through which to rotate $4x^2 - xy + y^2 + 2x + y - 1 = 0$ to eliminate the xy term.

(g) Plot the rotated curve above as we did in Solved Problem 5.

LABORATORY EXERCISE 18.1

Curve Fitting

Name ______________________ **Due Date** ______________

A small cannon is fired at a wall. The distance from the cannon to the wall, x meters, and the height of the spot on the wall where the projectile strikes, y meters, are recorded in the table:

x	3	3.5	4	4.5	5	5.5	6	6.5	7	7.5	8
y	1.3	1.4	1.6	1.5	1.5	1.4	1.5	1.5	1.4	1.2	1.3

1. We suspect that a quadratic $ax^2 + bx + c$ fits this data. Write down the matrix M whose generalized inverse must be found to do this.

2. Find the generalized inverse of the matrix M in Part 1.

3. Use generalized inverse of the matrix M in Part 1 to find the best least squares quadratic $ax^2 + bx + c$ that fits this data.

4. Check your result by using *DERIVE* 's FIT function.

5. Plot the data points and the quadratic you have found. Comment on your findings. Do they seem reasonable?

6. We might surmise from the graph that our cannon was not actually on the ground. Suppose that we *know* it's on the ground and therefore we really suspect ax^2+bx is the right formula. Repeat all the parts above in this problem to find and plot the best least squares quadratic $ax^2 + bx$ for the data.

LABORATORY EXERCISE 18.2

The QR Factorization

Name ______________________________ **Due Date** ____________________

Let $A = \begin{bmatrix} 2 & 2 & 1 & -2 \\ 2 & 2 & 0 & -2 \\ -6 & 5 & 5 & 13 \\ 4 & -3 & -1 & -7 \end{bmatrix}$.

1. Find the QR factorization, $A = QR$.

2. Show that the columns of Q are the result of applying the Gram-Schmidt algorithm to the columns of A. (See Chapter 12.)

3. Use the characteristic polynomial to obtain and approximate the eigenvalues of A.

4. Use 5, 10 and 30 iterations of the QR algorithm to approximate the eigenvalues of A and compare the results with your answer in Part 3.

5. Show that if A_n is the result of applying n iterations of the QR algorithm to A, then A is similar to A_n and hence they have the same eigenvalues. (Hint: No computer is needed here. Show first that for nonsingular matrices B and C, BC is similar to CB.)

Appendix I
DERIVE® Version 2.5 Reference

This appendix is not intended as a substitute for the *DERIVE User Manual*, which owners of authorized copies of *DERIVE* will have and should use. When this book was written, the authors used *DERIVE* version 2.5. If you are using an earlier version, you will notice a few differences. We have tried to mention them in the rare occasions when they might cause confusion. The most conspicuous new feature in version 2.5 is in the **Plot** command. See Section 13.

> *NOTE: DERIVE version 2.55 appeared after this book was composed. An internal graphics print capability was added and the line editing was enhanced so that one may now use the arrow keys to move the cursor on the author line.*

The *DERIVE* program, version 2.5, comes with a fine 246 page manual, but some people read software documentation only as a last resort. This appendix is a concise reference to the major features of *DERIVE* that arise often in the exercises in this book. Each topic in this appendix refers the reader to the *DERIVE User Manual* (Version 2.5) by section for detailed information. We encourage the interested reader to explore the *DERIVE* manual.

Special Functions and Symbols

Alt E or `#e`	the number e
Alt P or `pi`	the number π
Alt I or `#i`	the imaginary number i
inf	∞
−inf	$-\infty$
Alt Q	$\sqrt{}$

For the n by n identity matrix **Author** IDENTITY_MATRIX(n).

The escape key Esc will usually cancel a command or get you out of an undesirable situation.

Getting Started

DERIVE is driven by a menu at the bottom of the screen. You can select an item in two ways: (1) Type the capitalized letter in the name (for example, to select **soLve** press [L]) or (2) use the [Spacebar] to move through the menu and press [Enter] to select the highlighted item.

Entering arithmetic expressions involves customary syntax: addition (+), subtraction (−), division (/), exponents (∧), and multiplication (*) (however, multiplication does *not* require a *; $2x$ is the same as $2*x$).

Tutorial for Beginners

To enter an expression into *DERIVE* you almost always begin with the **Author** command. From the menu select [A] for **Author**, type **x^2 + 3x -4** on the *author line* and press [Enter] when done.

You will see that $x^2 + 3x - 4$ as it appears in expression 1 of Figure 0.1. (We have used *DERIVE* 's window-splitting capabilities to produce Figure 0.1. It is not necessary for you to do this, but if you want to learn how, consult Section 14.) Let's try some of the menu features: Hit [F] for **Factor**, then [Enter] [Enter], and note that the factored form $(x-1)(x+4)$ appears as expression 2. To expand it, choose the **Expand** command and then [Enter]. Now expression 2 has been expanded to its original form and appears as expression 3 of Figure 0.1.

The up and down arrow keys [↑] and [↓] move the highlight to other expressions. Commands apply to the highlighted quantity. With expression 3 highlighted, press [L] for **soLve** and then [Enter]. You will see that *DERIVE* lists the two solutions to the equation $x^2 + 3x - 4 = 0$ in expressions 4 and 5 of Figure 0.1. (When you ask *DERIVE* to **soLve** an expression that is not an equation, it attempts to find the zeros.)

Suppose that you want to evaluate the quadratic $x^2 + 3x - 4$ at $x = 2$. Highlight it first; then use the commands **Manage Substitute** [Enter]. Type 2 for x at the prompt and then [Enter]. The result, $2^2 + 3\ 2 - 4$, appears in expression 6 . **Simplify** it to get the answer, 6, in expression 7.

In mathematics an equal sign may denote an equation to be solved (such as $x^2 + 3x - 4 = 0$) or a logical statement, which might be true or false (such as $4 = 2 + 2$), or a definition (such as $a = 2$). *DERIVE* uses the equal sign for the first two cases, but for the third, the definition, it is necessary to use "colon equal." Thus, to assign a the value 2 you would **Author a:=2**. *DERIVE* now knows that a represents the number 2. If you **Author a^2** and **Simplify** , *DERIVE* will return the value 4 in expression 10 of Figure 0.1. This assignment will continue in force until you specifically change it. To "un-define a, **Author a:=** .

The "colon equal" can also be used to define functions. See Section 12 for more details.

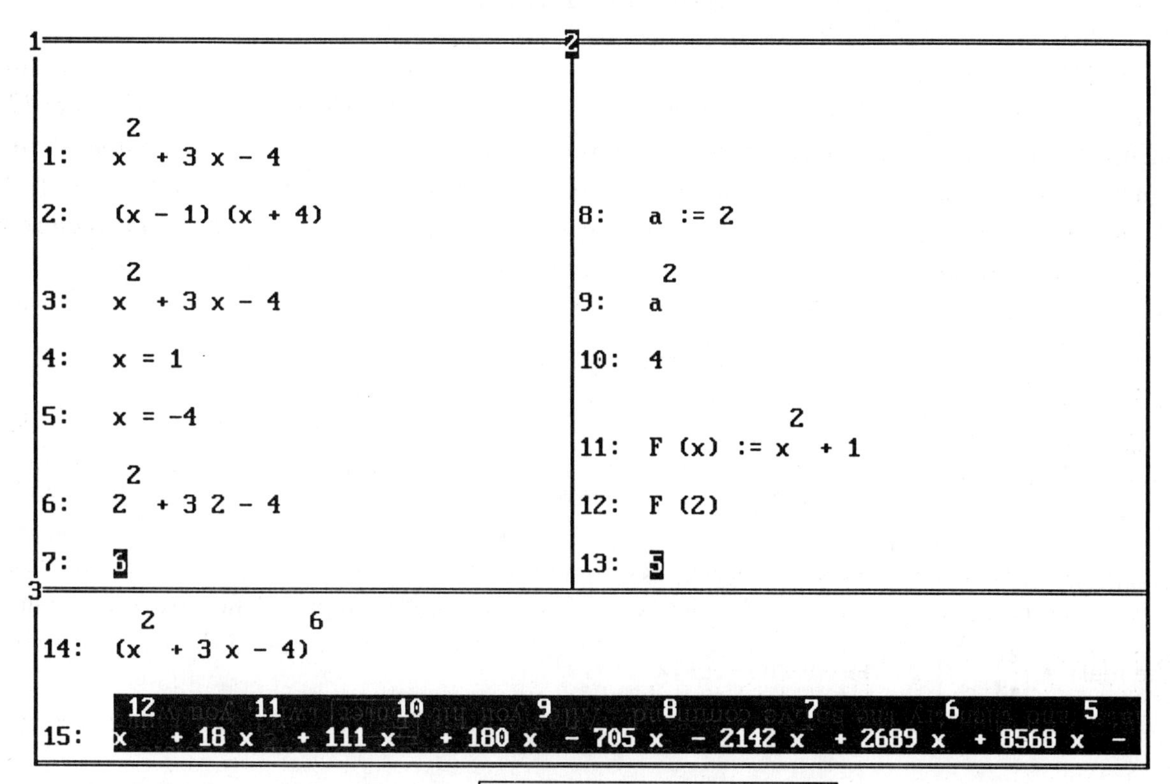

Figure 0.1: **A tutorial for beginners**

You may always refer to an expression by **#** followed by its expression number. If you want the sixth power of the quadratic x^2+3x-4 in expression 1, you just **Author #1^6** and $(x^2+3x-4)^6$ appears. (See expression 14 of Figure 0.1.) In this way complex expressions may be authored piece by piece and then assembled by expression number. To expand it, use **Expand**. Notice that expression 15 is so long that it runs off the screen. To see it all, use Ctrl → and Ctrl ← to scroll back and forth. (If you print such an expression, it will "wrap around" on the printer.)

There is an alternative method for calculating $(x^2+3x-4)^6$. The F4 key brings whatever expression is highlighted down to the *Author line* enclosed in parentheses. Thus if you highlight expression 1 and **Author** F4^6, *DERIVE* will produce expression 14 as before. This method is necessary if you wish to operate on a portion of an expression rather than the whole thing.

Section 1 Entering Matrices and Vectors
DERIVE Manual Chapter 8.

To enter a matrix into *DERIVE*, issue the **Declare Matrix** commands. See Figure 0.2. *DERIVE* will ask you for the size of the matrix you wish to enter. When you've given the number of rows and columns, you'll be prompted for the entries.

Let's enter a 2 by 2 matrix. Change the default **Rows: 3** to 2. Press the Tab key to move the cursor to **Columns:**, change the 3 to 2, and press Enter. See Figure 0.2. *DERIVE* will now prompt you to enter the elements of the matrix. Let's enter them as 1, 2, 3, 4 in order. Notice that the (i, j) position of the entry appears on the status line at the bottom of the screen as you proceed. See Figure 0.3. Once you have finished, *DERIVE* will display the matrix $\begin{bmatrix} 1 & 2 \\ 3 & 4 \end{bmatrix}$ as in Figure 0.4.

Entering a vector is done in exactly the same way by using the **Declare Vector** commands. If the vector is small (say, [1, 2]) you may simply **Author** [1, 2], but you *must* use square brackets. <1, 2> or $(1, 2)$ will not work. See Figure 0.4.

To *DERIVE*, a matrix is a vector whose entries are vectors. For example, you could enter the matrix $\begin{bmatrix} 1 & 2 \\ 3 & 4 \end{bmatrix}$ by **Authoring** `[[1,2],[3,4]]`.

```
DECLARE MATRIX: Rows: 2        Columns: 2

Enter number of columns
                                   Free:100%          Derive Algebra
```

Figure 0.2: Setting the size of a matrix

```
MATRIX element: 2

Enter matrix element (1,2)
                                   Free:100%          Derive Algebra
```

Figure 0.3: Entering elements of a matrix

```
1:   [ 1  2 ]
     [ 3  4 ]

2:   [1, 2]
```

Figure 0.4: **A matrix and a vector**

What if I make a mistake? If you find that you made a mistake after the matrix appears on the screen and want to change an entry, press A for **Author** and then F3 to bring the matrix to the author line for editing. See section 8 for details on this feature.

If you realize that you made a mistake *before* the matrix appears on the screen, you cannot back up. You must either finish entering the elements and edit the result or press Esc to start over again.

Section 2 Uppercase and Lowercase Letters and Multiletter Names *DERIVE* Manual Section 4.1

Although you may refer to a matrix by its line number, you may also want to assign it a name such as $A := \begin{bmatrix} 1 & 2 \\ 3 & 4 \end{bmatrix}$. When you do so you may find that *DERIVE* changes the letter "A" to lowercase, but you can instruct it to be case-sensitive. Since most books use upper-case letters for matrices, you may want to do this. However, we have chosen not to because it causes syntax errors when you forget to type predefined functions such as IDENTITY_MATRIX in capital letters.

If you want to designate things with capital letters or with names having more than one letter, use the commands **Options Input**. *DERIVE* will present you with the options displayed in Figure 0.5.

If you change to `Word` by pressing W, you may designate a variable with a name having more than one letter. To use capital letters, Tab to `Case:` and press S for `Sensitive`.

Warning: If "word mode" is selected, variables must be separated by a space. For example, you will have to **Author** `sin x` and not `sinx`. If "case-sensitive mode" is selected, pre-defined functions such as SIN and LN *must* be entered in uppercase.

```
OPTIONS INPUT: Mode: Character Word  Case:(Insensitive)Sensitive

Select input mode
                                        Free:100%            Derive Algebra
```

Figure 0.5: Changing the input mode

Section 3 Solving Systems of Equations
DERIVE Manual Section 4.15

Systems of linear equations must appear in a list separated by commas and enclosed by square brackets. If the system is *homogeneous,* that is, the constant terms are all zeros, then it is not necessary to type the "=0." Asking *DERIVE* to **soLve** `[x+y,2x-y]` and `[x+y=0,2x-y=0]` gives the same result, as can be seen from expressions 2 and 4 of Figure 0.6.

When entering a large system of equations it may be best to **Author** each equation so that each equation appears as a separate expression. Consider the following system of equations.

$$\begin{aligned} x+y+z &= 3 \\ x-y+2z &= 4 \\ 2x-3y-z &= 5 \end{aligned}$$

In Figure 0.6 we have authored each equation separately as expressions 5, 6, and 7. The equations were assembled in expression 8 using `[#5,#6,#7]`. **soLve** produces the result in expression 9. The advantage of this method is that it avoids having to edit large expressions.

Sometimes you may encounter the matrix form of a system such as $\begin{bmatrix} 1 & 2 \\ 3 & 4 \end{bmatrix}\begin{bmatrix} x \\ y \end{bmatrix} = \begin{bmatrix} 6 \\ 5 \end{bmatrix}$. See section 6 for hints in this case.

Section 4 The Distinction Between Matrices and Vectors:
An Important Warning

Most texts do not distinguish between the 1 by 3 matrix [1 2 3] and the vector [1, 2, 3]. *DERIVE* sees them as entirely different objects and it is important to understand this to use the software. To see the difference, we'll enter the matrix [1 2] and the vector [1,2].

Use **Declare Matrix**, set **Rows: 1**, press the Tab key to move the cursor to Columns, and set it to 2. Enter the values 1 and 2. The result appears as expression 1 of Figure 0.7.

```
1:   [x + y, 2 x - y]

2:   [x = 0, y = 0]

3:   [x + y = 0, 2 x - y = 0]

4:   [x = 0, y = 0]

5:   x + y + z = 3

6:   x - y + 2 z = 4

7:   2 x - 3 y - z = 5

8:   [x + y + z = 3, x - y + 2 z = 4, 2 x - 3 y - z = 5]

9:   [x = 28/11, y = - 2/11, z = 7/11]
```

$$\left[x = \frac{28}{11},\ y = -\frac{2}{11},\ z = \frac{7}{11}\right]$$

Figure 0.6: **Solving systems of equations**

Use **Declare Vector**, set **Dimension: 2** and enter the values 1 and 2. The result appears as expression 2 of Figure 0.7.

If you examine the results carefully, you will see that the matrix in expression 1 has no comma separating the 1 and the 2, while the vector in expression 2 does have a comma. It's instructive to look at these on the author line. Highlight each of them and press F3 to do this. You will see that, on the author line, the vector appears as [1, 2] while the matrix appears as [[1, 2]].

The general idea is that a matrix is a vector whose entries are vectors of equal dimensions. Here's an important example: If you **Author** and **Simplify** `#1'`, *DERIVE* will return the transpose of the matrix as in expression 4 of Figure 0.7. (*DERIVE*'s syntax for the transpose is '; that's (`), not an apostrophe.) If, on the other hand, you **Author** and **Simplify** `#2'`, *DERIVE* will simply return the input as in expression 6. This indicates that *DERIVE* is willing to calculate transposes of matrices but not of vectors.

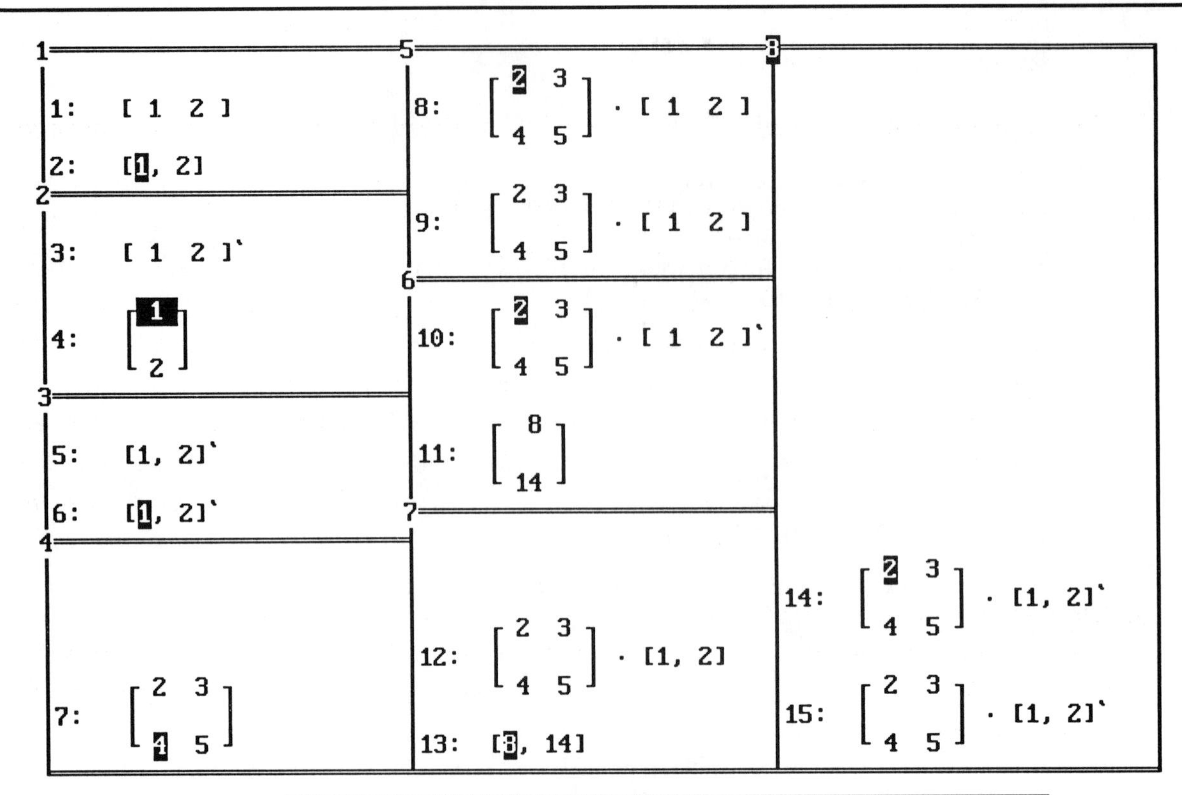

Figure 0.7: **DERIVE 's distinction between vectors and matrices**

Section 5 Vector and Matrix Operations

DERIVE Manual Sections 8.4 and 8.5

1. *DERIVE* 's syntax for matrix and vector *addition* (+) and *subtraction* (−) are customary.

2. Taking a *power* of a square matrix uses (∧) that is, `A^2`.

3. The *dot product* (or scalar product) of two vectors as well as *matrix multiplication* must be designated by a period (.) (e.g., `A.B`) and, unlike numerical multiplication, you cannot use juxtaposition (AB) or ∗ (A∗B).

   There are two situations to be careful about.

   (a) If A and B are matrices and you **Author** and **Simplify** the expression `A^2.B`, *DERIVE* will not return the product of A^2 and B. The difficulty is that when *DERIVE* sees a period immediately following an integer, it interprets the period as a decimal point rather than the symbol for matrix multiplication. You can avoid this by using parentheses or by leaving a space after the 2. Thus, either `(A^2).B` or `A^2 .B` will produce the correct answer.

(b) Suppose we want to multiply the vector [1, 2] by the matrix $\begin{bmatrix} 1 & 2 \\ 3 & 4 \end{bmatrix}$. Many books will treat [1,2] as a one by two matrix and write either $\begin{bmatrix} 1 & 2 \\ 3 & 4 \end{bmatrix}$ [1,2]‘ or $\begin{bmatrix} 1 & 2 \\ 3 & 4 \end{bmatrix}\begin{bmatrix} 1 \\ 2 \end{bmatrix}$. The way you indicate this product in *DERIVE* depends on whether you are dealing with the *matrix* [1 2] or the *vector*, [1, 2]. (The period (.) is required in either case.) In window 5 of Figure 0.7 *DERIVE* cannot calculate the product because the matrices are not appropriately sized. In window 6 the product of the two matrices is correctly calculated using the transpose. In window 7 the product of the matrix and the vector is correctly calculated. *(Notice that the output here is a vector while the output in window 6 is a matrix.)* In window 8 the product is not calculated because *DERIVE* does not think that the vector [1, 2] has a transpose.

4. **Elementary Row Operations**. The three elementary row operations must be loaded from the VECTOR.MTH file that comes with the *DERIVE* program. To do this, use **Transfer Merge** and then type VECTOR. You will see many functions appear on the screen, among them the following:

   (a) SCALE_ELEMENT(A,r,s) multiplies row r of matrix A by the number s.

   (b) SUBTRACT_ELEMENTS(A,i,j,s) subtracts s times row j of matrix A from row i.

   (c) SWAP_ELEMENTS(A,i,j) interchanges rows i and j of matrix A.

5. Matrix *inversion* is designated by raising it to the power -1 e.g., `A^-1`.

6. The *cross product* of two vectors is denoted by CROSS *e.g.,* `CROSS([1,2,3],[4,5,6])`.

7. The *length* or *norm* of a vector is denoted by vertical bars or absolute value. Thus, $|[1, 2, 3]|$ =ABS[1,2,3] $= \sqrt{14}$. (If you have an early version this may not work. Define your own function LEN`(x):=sqrt(x.x)`.)

8. The *transpose* of a matrix A is denoted by A‘. (That's (‘), not an apostrophe.) *Warning: DERIVE does not recognize the transpose of a vector, just a matrix.*

9. The *characteristic polynomial* of a matrix A is denoted by CHARPOLY(A).

10. The *determinant* of a matrix A is denoted by DET(A).

Section 6 More on Solving Systems

Sometimes you may encounter the matrix form of a system such as $\begin{bmatrix} 1 & 2 \\ 3 & 4 \end{bmatrix} \begin{bmatrix} x \\ y \end{bmatrix} = \begin{bmatrix} 6 \\ 5 \end{bmatrix}$. If you enter it like this, `A.[x,y]=[6,5]`, then you can just **Simplify** it and it will appear as $[x+2y=6,\ 3x+4y=5]$, which is a vector of equations that can then be **soLved**. (This may not work in earlier versions of *DERIVE*.)

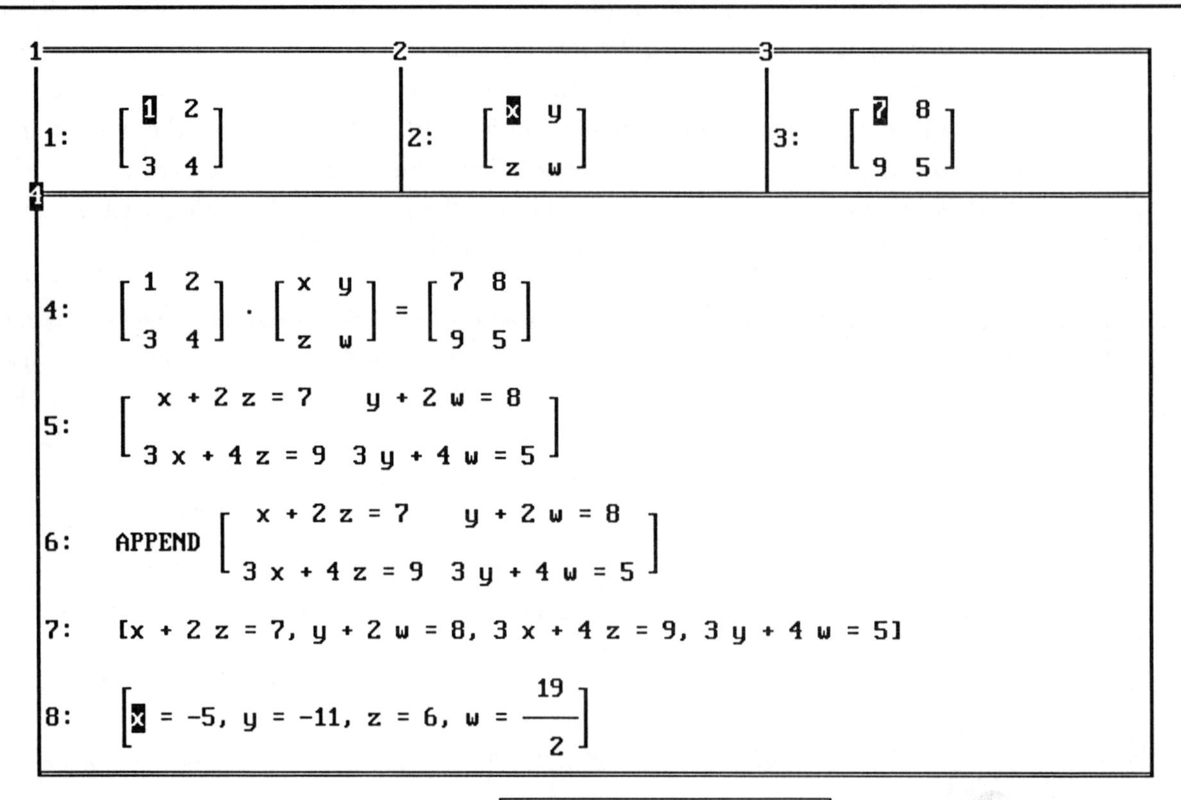

Figure 0.8: **A matrix of equations**

Occasionally we encounter a *matrix* of equations. Suppose we must solve $\begin{bmatrix} 1 & 2 \\ 3 & 4 \end{bmatrix} \cdot \begin{bmatrix} x & y \\ z & w \end{bmatrix} = \begin{bmatrix} 7 & 8 \\ 9 & 5 \end{bmatrix}$. *DERIVE* will not solve the equation as written, or even after it has been simplified. We need to list the four equations as the elements of a vector, which is best done with the APPEND function. These three matrices appear as expressions 1, 2, and 3 of Figure 0.8. We build the appropriate equation in expression 4 using **#1.#2=#3**. **Simplify** to obtain the matrix of equations in expression 5. (This may not work in earlier versions of *DERIVE*.) **Author**

append(#5) and **Simplify** to transform the matrix of equations into a vector of equations as in expression 7. **soLve** produces the solution in expression 8 of Figure 0.8.

If you are using an early version of *DERIVE* you will notice that APPEND is not defined. We have provided a definition in Appendix II.

General Features

Section 7 Moving Around
DERIVE Manual Section 3.3.

When you have a long session with lots of expressions, it may become awkward to use the arrow keys to move up and down the screen. Page Up and Page Down move in bigger jumps. You can use Home and End to go between the top and bottom of the stack. To go to a given expression number, use the command **Jump** followed by the number of the expression.

Section 8 Line Editing
DERIVE Manual Section 2.4

NOTE: DERIVE versions 2.55 and 2.55C appeared after this book was composed. In version 2.55C the line editing was enhanced so that one may now use the arrow keys to move the cursor on the author line.

```
1:  [ 1  66 ]
    [ 3   4 ]

AUTHOR expression: [[1, 66], [3, 4]]

Enter expression
User                                   Free:100%            Derive Algebra
```

Figure 0.9: **A matrix on the Author line for editing**

Suppose you want to change or edit an expression in the window. Highlight it by using the arrow keys ↑ ↓, press A for **Author**, and then press F3. You should see the expression

```
1:  [ 1  66 ]
    [ 3   4 ]

AUTHOR expression: [[1, 2], [3, 4]]

Enter expression
User                                   Free:100%              Derive Algebra
```

Figure 0.10: **"66" changed to "2" on the Author line**

```
1:  [ 1  66 ]
    [ 3   4 ]

2:  [ 1  2 ]
    [ 3  4 ]
```

Figure 0.11: **Editing completed**

appear on the author line where you can edit it. F4 will bring an expression to the author line enclosed in parentheses.

Ctrl S moves the cursor left on the author line and Ctrl D moves it right. You will observe that the arrow keys move the highlight in the algebra window and do not affect the cursor on the author line. As usual, the Insert key toggles between typeover and insert modes; the Delete and backspace keys erase. When you are finished, hit Enter and the new expression appears on the bottom of the list of expressions.

This procedure is used to edit an entry of a matrix in Figures 0.9, 0.10 and 0.11.

Section 9 Approximations and Precision
DERIVE Manual Section 3.8

If you want *DERIVE* to *always* give approximate answers, choose **Options Precision Approximate.** To make it give exact answers choose **Options Precision Exact**. Don't forget to change this setting at the appropriate time.

When *DERIVE* is set on **Exact**, you may still get decimal approximations to numbers using [X] for **approX**. For example, **Author** `2^(.5)`. After it appears in the window, press [X] [Enter]. To set the number of decimal places in the approximation, use the **Options Precision** commands. The [Tab] key jumps to the number where you indicate how many decimal places you would like.

Section 10 Solving Equations Exactly
DERIVE Manual Section 4.14

Author the equation you wish to solve, highlight it, and use the **soLve** command. If the precision is set on **Exact** (see Section 9 above), *DERIVE* will attempt to give the exact roots. **Author** and **soLve** the expression `2x^4+9x^3-27x^2=22x-48`. *DERIVE* will find the four solutions in expressions 2, 3, 4, and 5 of Figure 0.12.

If you ask *DERIVE* to **soLve** an expression that is not an equation, inequality or other relation, it will try to find the zeros of the expression. For example the expression $x^2 - 2$ is **soLved** in expressions 6 through 8 of Figure 0.12.

If *DERIVE* is able to determine that there are no solutions, it will beep and say "No solutions found." In expression 9 of Figure 0.12 we have asked *DERIVE* to **soLve** $x + 1 = x - 1$. Notice the message on the status line at the bottom.

Caution: If you ask *DERIVE* to solve $x^2 + 4 = 0$, you may expect "No solutions found," but it will give the complex number solutions $2i$ and $-2i$.

When a solution exists but cannot be found exactly *DERIVE* simply returns the equation. For example, try $x^5 + 5x^2 - 1$. In this case it will be necessary to approximate the solutions. See the next section.

Section 11 Solving Equations Approximately
DERIVE Manual Section 4.14

Author the equation you wish to solve, highlight it, and use the **soLve** command. If you ask *DERIVE* to **soLve** an expression that is not an equation, inequality, or other relation, it will try to find the zeros of the expression. If the precision is set on **Approximate** (see Section

```
         4       3        2
1:    2 x  + 9 x  - 27 x  = 22 x - 48

              3
2:    x = - ───
              2

3:    x = 2

              √57     5
4:    x = - ───── - ───
               2      2

            √57     5
5:    x = ───── - ───
             2      2

       2
6:    x  - 2

7:    x = - √2

8:    x = √2

9:    x + 1 = x - 1
COMMAND: Author Build Calculus Declare Expand Factor Help Jump soLve Manage
         Options Plot Quit Remove Simplify Transfer moVe Window approX
No solutions found
User                                   Free:100%                 Derive Algebra
```

Figure 0.12: Exact solutions of equations

9), *DERIVE* will apply a technique called *the bisection method* to find an approximate answer. To do this, it must begin with an interval in which to search for a root. The smaller the interval, the faster *DERIVE* will find an answer.

Let's try $x^2 - 2$. **Author** `x^2 - 2`, execute the commands **Options Precision Approximate**, and then try the **soLve** command. After you hit Enter twice, you will see on the command line at the bottom of the screen SOLVE: Lower: -10 Upper: 10. *DERIVE* is offering to search for a solution of the equation on the interval $[-10, 10]$. This is the default interval. If you press Enter, *DERIVE* returns 1.41422. If you use **soLve** again but change the interval to -10 to 0, you will get -1.41422. (You move to the second endpoint with the Tab key.)

If you ask *DERIVE* to **soLve** $x^5 + 5x^2 - 1$, it will find no exact roots. This is indicated by *DERIVE* 's response in expression 2 of Figure 0.13. But there are approximate ones. How do you know what interval or intervals to search on? The answer is to **Plot** it and use the graph

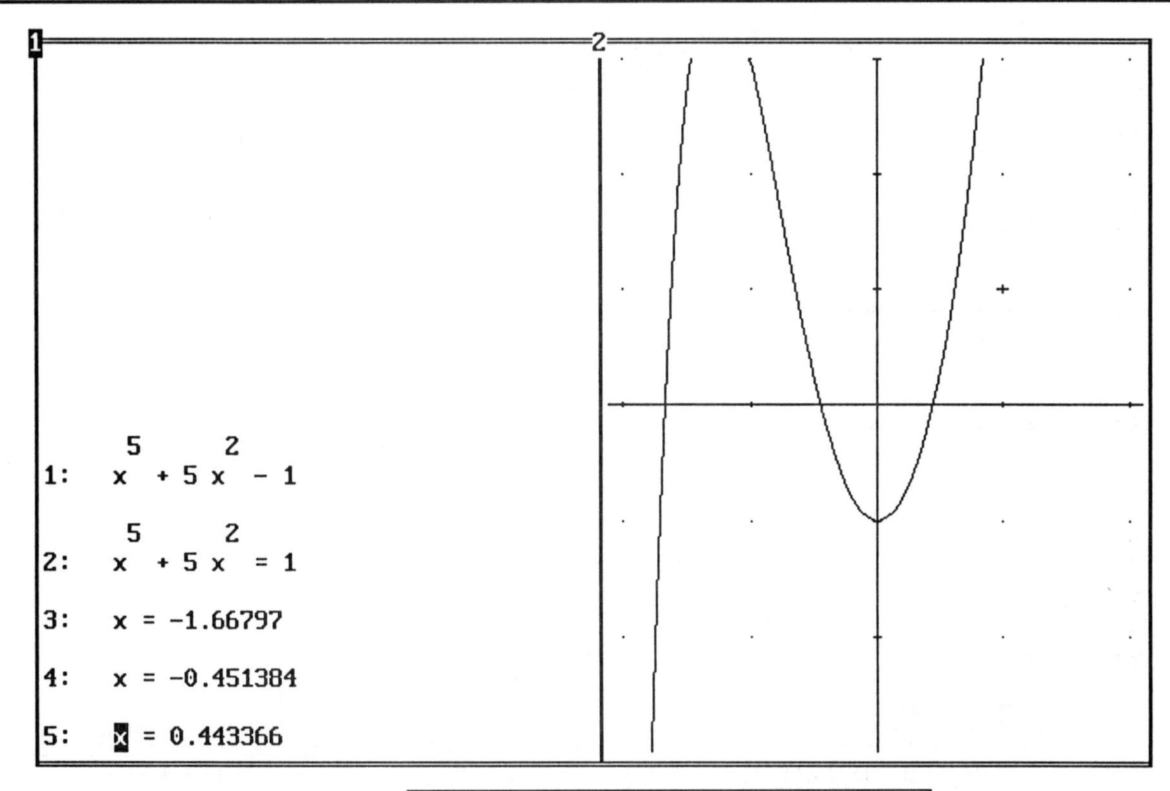

Figure 0.13: **Approximate solutions of equations**

to estimate these intervals. (See Section 13.) In window 2 of Figure 0.13 we see that the graph of $x^5 + 5x^2 - 1$ crosses the x-axis once on each of the intervals $(-2, -1)$, $(-1, 0)$, and $(0, 1)$. Expressions 3, 4 and 5 of Figure 0.13 shows the results of **soLving** on each of these intervals.

Don't forget to reset precision to **Exact** at the appropriate time. If you try to **soLve** $x + a$ for x in approximate mode, *DERIVE* will beep and say "No solutions found."

Section 12 Functions
DERIVE Manual Section 4.12

To define a function, for example, $f(x) = 1 + x^2$, **Author** **`f(x):=1+x^2`** Enter. (Be sure to put the colon before the equal sign.) From now on *DERIVE* will know that $f(x)$ is this function. If you want to evaluate the function when $x = 3 + t$, **Author** **`f(3+t)`**. When it appears in the window, hit S for **Simplify**.

If you want to redefine f, repeat the process above. The last definition you **Author** is the one *DERIVE* accepts. If you no longer want f to be a function, **Author** **`f:=`** and then Enter.

Section 13 Plotting Graphs
DERIVE Manual Section 5.2

To plot a graph, highlight the expression to be plotted, press P for **Plot**, and you are given three choices **Beside Under Overlay** followed by a choice of where the split will be. (In versions prior to 2.5 you would be put in an overlay plot window automatically.) Press P for **Plot** again and the highlighted expression will be graphed.

Suppose that you want to plot another expression. Hit A to return to the **Algebra** window and highlight the new expression for plotting. If you now hit P for **Plot** twice as before, *both* graphs will be plotted. To get rid of the original graph, use the **Delete First** command; more about this later.

Plot Features

- **Algebra** returns to the algebra screen from the plot screen.
- Esc stops plotting. P for **Plot** starts it.
- The four arrow keys move the small cross on the screen. Notice that the coordinates of the cross appear in the lower left corner. Ctrl ← Ctrl → Page Up and Page Down move the cross in bigger jumps.
- **Center** centers the graph on the cross.
- **Scale** lets you change the scales on the x-axis and y axis.
- F9 zooms in, F10 zooms out.
- *Plotting several graphs at once*: If you want to plot three or more graphs, **Author** them in a list separated by commas and enclosed in square brackets. For example, `[sin x, sin(2x), sin(3x)]`. When you **Plot** this list, all three graphs will be drawn in order. If you want just two, for example $\sin x$ and $\sin(2x)$, you may **Author** and **Plot** `[sin x, sin(2x), ?]`. If you omit the question mark and just plot $[\sin x, \sin(2x)]$, *DERIVE* will interpret it as a single parametric curve. Try it.
- *Parametric plots*: If you ask *DERIVE* to **Plot** a vector containing exactly two functions such as $[\cos x, \sin x]$, *DERIVE* will assume that you are asking for a parametric plot and will ask for a *parameter domain*. You will see the prompt PLOT:Min:−3.1416 Max: 3.1416 at the bottom of the screen. This is where you set the interval for x. If you wish to change the interval to $[-1, 1]$, proceed as follows. Delete the Min: offering and type `-1`. Use the Tab key to move to the Max: setting. Delete it and type 1. Now hit Enter and the graph will be drawn.

- *Plotting a graph over a specific interval.* Suppose that you want to plot x^2 over the interval $[-1, 0.5]$. **Author** `[x, x^2]` and **Plot** it as described in *parametric plots* above, changing the parameter interval to -1 and 0.5.

- *Plotting individual points.* If you want to plot several points on the screen, for example, $(1,2)$, $(2,2)$, and $(3,-1)$, **Author** the list `[[1,2],[2,2],[3,-1]]` and hit P for **Plot** twice.

Five Common Questions About Graphics and Plotting

- *The graph is just a bunch of little dots. What's wrong?*

  You are not in graphics mode. Use the **Options Display** commands in the plot window to change from Text to Graphics and also to set the correct graphics adapter. (See Section 5.1 of your *DERIVE* Manual.) After this, F5 will flip between text and graphics modes if required. (You may save your two most recent screen modes for later sessions with the **Transfer Save State** commands.)

- *I tried to plot a new graph and I got another one I did earlier. How do I get rid of it?*

  DERIVE saves all your 2D plots. If you plot a graph and then go to the algebra screen and plot another, the first will be graphed followed by the second. Use the command **Delete** and you see the prompt: **DELETE All Butlast First Last**. These four choices are basically self-explanatory, but if you just want the graph you last asked to plot, press A for **All** and then P for **Plot**.

- *I tried to plot something and DERIVE beeped at me. In the lower left corner of the plot screen I saw the message "Cannot do implicit plots." What's wrong?*

  You probably are trying to plot an equation such as $x^2 + y^2 = 1$ or $y^2 = x^2$. *DERIVE* plots *expressions* but not equations. There is an exception: *DERIVE* will plot an equation of the form $y = f(x)$ as if it were $f(x)$ alone. To plot an equation such as $x^2 + y^2 = 1$, you may solve for y and plot the two solutions $\pm\sqrt{1 - x^2}$ together.

- *How do I plot a vertical line?*

  To plot the line $x = 3$, for example, **Author** `[3,y]` and **Plot** it. This is a parametric plot and *DERIVE* will ask for a parameter interval. (See "Plotting a graph over a specific interval" above.) The defaults are -3.1416 and 3.1416. If you hit Enter to accept them, you will get a plot of the line $x = 3$ from $y = -3.1416$ to $y = 3.1416$.

- *How do I print a graph?*

  DERIVE version 2.54 has no internal graphics printing capability; however, GRAPHICS.COM in DOS 5.0 works well. Consult your MicroSoft DOS 5 manual under Graphics for a list of supported printers and more details. You can print text from *DERIVE* with the **Transfer Print** command.

  NOTE: DERIVE versions 2.55 and 2.55C appeared after this book was composed. In version 2.55 an internal graphics print capability was installed.

  There are many "graphics dump" and "graphics capture" programs available. For a list of several of these, please see the *DERIVE User Manual*, version 2, page 223. We used RAINDROP[tm] and an HP LaserJet ® III printer for the figures in this book.

Section 14 Splitting the screen into windows
DERIVE Manual Section 5.7

You may split the screen into windows to view different sessions at once. The most frequent use of this feature is to view algebra and graphs at the same time. In version 2.5 this is automatic when you **Plot**; you are given three choices **Beside Under Overlay**. In earlier versions this is not the case and the splitting must be done manually. You can split into left and right windows using the **Window Split Vertical** commands or upper and lower ones using the **Window Split Horizontal** commands. In any case you may want to have two sessions displayed at once. We'll present an example:

1. **Author x^2.**
2. Issue the **Window Split Vertical** commands, then [Enter], and you will see two windows numbered 1 and 2. Press the [F1] key to flip between the windows.
3. With window 2 highlighted, press [P] for **Plot** twice and the graph of x^2 will appear in window number 2.
4. Press [F1] to move to window 1.
5. **Author sin x.** Now press [P] for **Plot** twice.

   Comment: You do not have to move to window 2. The graph will appear there automatically. However, it is a good idea to use the commands **Window Designate 2D** to designate window 2 as a 2D plot window. This saves memory taken up by the duplicate algebra window.

6. If you want to return to a full screen, highlight window 2 with the F1 key and issue the **Window Close** commands. You will have to do this twice, first to close the plot window and second to close the algebra window. (Answer "yes" to the prompt "Abandon expressions (Y/N)?")

Section 15 Substituting into an expression
DERIVE Manual Section 4.8

You often want to plug a value of x into a function or substitute one expression for another. To do this use the commands **Manage Substitute.** Here are some examples. We will use the expressions $(x+1)^2 + \sin(x+1)$ and x^3 so **Author** **(x+1)^2 + sin(x+1)** and **x^3**. Refer to Figure 0.14.

```
            2
1:   (x + 1)  + SIN (x + 1)

      3
2:   x

            2
3:   (3 + 1)  + SIN (3 + 1)

4:   SIN (4) + 16

       3      2          3
5:   (x  + 1)  + SIN (x  + 1)

      2
6:   t  + SIN (t)
```

Figure 0.14: Substituting in expressions

Example A. Substitute the number 3 into expression 1 for x.

1. Highlight expression 1 and use **Manage Substitute.**
2. Acknowledge the prompt "Substitute expression #1" with [Enter].
3. Next you are asked if the variable that you want to replace is x. Type 3 [Enter].
4. You should now see expression 3 of Figure 0.14.
5. **Simplify** expression 3 to get the desired result as expression 4.

Example B. Substitute x^3 (expression 2) into expression 1 for x. (This illustrates how you may substitute for x in an existing expression. This is handy when you want to substitute a complicated expression without retyping it.)

1. Again highlight expression 1 and use **Manage Substitute.**
2. Acknowledge the prompt "Substitute expression #1" with [Enter].
3. Next you are asked if the variable that you want to replace is x. Type **#2** [Enter].
4. You should now see expression 5 of Figure 0.14.

Example C. Substitute t into expression 1 for $x + 1$ *everywhere it occurs in the formula.*

1. First highlight expression 1. Now, we want to highlight the subexpression to be replaced. To do this, use the arrow keys in the following order: [←] [↓]. (Notice the placement of the highlight in expression 1 of Figure 0.14.)
2. Use **Manage Substitute** and you see the prompt MANAGE SUBSTITUTE value: . Type t and [Enter].
3. You should now see expression 6 of Figure 0.14. Notice that *both* occurrences of $x+1$ have been replaced by t.

Section 16 Complex Numbers
DERIVE Manual Section 4.5

DERIVE actually works in the field of complex numbers. Therefore, any of the examples in this book may be modified to involve complex numbers. The main thing to keep in mind is that the complex number i must be entered in *DERIVE* as [Alt I]. (*DERIVE* treats the letter i as just another variable.)

Here are some very simple examples for illustration:

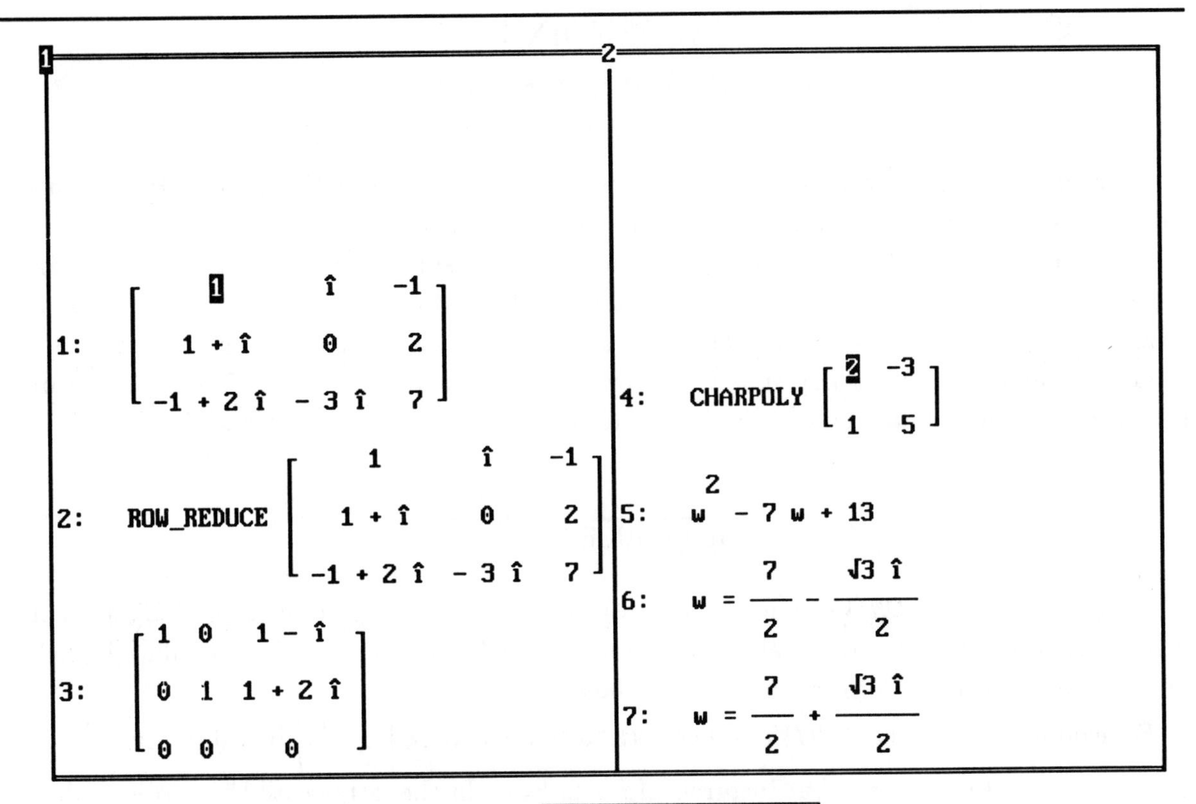

Figure 0.15: **Complex numbers**

Use **Declare Matrix** to enter $\begin{bmatrix} 1 & i & -1 \\ 1+i & 0 & 2 \\ -1+2i & -3i & 7 \end{bmatrix}$ as in expression 1 of Figure 0.15. When you ROW_REDUCE it you get expression 2.

Similarly, *DERIVE* will find determinants and inverses of matrices with complex entries.

Author charpoly([[2, -3],[1, 5]]) and **Simplify** to get $w^2 - 7w + 13$ as in expressions 4 and 5 of Figure 0.15. When you **soLve** you get the two complex eigenvalues $w = \dfrac{7 - \sqrt{3}i}{2}$ and $w = \dfrac{7 + \sqrt{3}i}{2}$.

APPENDIX II

Optional User-Defined Auxiliary Files

There may be some specialized things we want to do that *DERIVE* does not provide, so we have to design the code ourselves. That is the purpose of this appendix.

Start with a blank screen and follow the instructions carefully. Type everything *exactly* as it appears: The colon is essential in := and watch for the transpose symbol ‘ which is usually located on the same key as the symbol ~ on a standard keyboard. It is *not* an apostrophe (’).

Loading files and doing things in the right order are important. Save your work using **Transfer Save Derive** followed by the file name and recall it using **Transfer Load Utility**.

The REDUCE.MTH File

Discussion: The *DERIVE* function ROW_REDUCE will return the reduced echelon form of a matrix. However, if there are variables in the matrix, *DERIVE* might produce a reduced echelon matrix that is incorrect for certain values of those variables.

For example, if you ask *DERIVE* to ROW_REDUCE the matrix $\begin{bmatrix} 1 & a \\ 2 & b \end{bmatrix}$ it will return $\begin{bmatrix} 1 & 0 \\ 0 & 1 \end{bmatrix}$. This is correct if $b - 2a \neq 0$, but incorrect if $b - 2a = 0$. In the latter case the correct reduced echelon form is $\begin{bmatrix} 1 & a \\ 0 & 0 \end{bmatrix}$. Thus, *DERIVE* treats the variables as if they do not have exceptional values that may lead to division by zero, and this can lead to difficulties in solving some systems of equations. If we ask *DERIVE* to ROW_REDUCE $\begin{bmatrix} 1 & 1 & a \\ 2 & 2 & b \end{bmatrix}$, it will return $\begin{bmatrix} 1 & 1 & 0 \\ 0 & 0 & 1 \end{bmatrix}$, which implies that the system of equations

$$\begin{aligned} x + y &= a \\ 2x + 2y &= b \end{aligned}$$

has no solution. In fact, if $a = 1$ and $b = 2$, there are infinitely many solutions.

For this reason, we often want to apply to a matrix the row operations required to reduce only the first few columns. This is the case when doing Gauss-Jordan elimination: We want to apply to an augmented matrix the row operations required to reduce only the *coefficient matrix*. Therefore, we will find it useful to create a function of our own called REDUCE that will accomplish this.

If A is a matrix, the user-function REDUCE(A, n) described below will apply to A the row operations required to put its first n columns into reduced echelon form.

Begin with a clear screen and **Author** the following expressions *exactly* as they appear. When you are finished, save the file using **Transfer Save Derive** REDUCE.

```
d_(a):=dimension(a)
i_(n):=identity_matrix(n)
e_(a,i):=element(a,i)
r_r(a,b):=row_reduce(a,b)
v_(a,n):=vector(e_(a,i),i,n)'
L_(a,n):=vector(e_(a,i),i,n+1,d_(a))'
m_(a,n):=L_(r_r(v_(a',n),i_(d_(a)))',n)
reduce(a,n):=m_(a,n).a
```

Be sure to save this file using **Transfer Save Derive** REDUCE.

The APPEND.MTH File

This file is only for those who have a version of DERIVE prior to 2.5. The APPEND *function is internal to version 2.5.* The APPEND function is used and discussed throughout the book, and we will not do so here.

When you need to use APPEND, be sure to use **Transfer Load Utility** VECTOR first and then **Transfer Load Utility** APPEND.

Begin with a clear screen and **Transfer Load Utility** VECTOR. When it has loaded, **Author** the following expressions *exactly* as they appear.

```
H(a,v,n):=IF(n<=DIMENSION(a), H(a, APPEND_VECTORS(v, ELEMENT(a,n)), n+1), v)

APPEND(a):=H(a, ELEMENT(a,1), 2)
```

Be sure to save this file using **Transfer Save Derive** APPEND.

The RANDOM.MTH File.

Discussion: Sometimes we might just like to have a "random" matrix with integer entries that we can use for an example. The following function RAND_MAT(n,m) will generate an n by m matrix with integer entries between -10 and 10.

Begin with a clear screen and **Author** the following expressions *exactly* as they appear. When you are finished, save the file using **Transfer Save Derive** RANDOM.

```
FLOOR(10(1- 2RANDOM(1)))

VECTOR(#1, i, 1, m)

RAND_MAT(n,m):=VECTOR(#2, j, 1, n)
```

When you need RANDOM, first use **Transfer Load Utility** RANDOM. Now, if you want a 4 by 2 matrix, for example, **Author** RAND_MAT(4,2) and **Simplify**.

The LU.MTH File

This code produces the LU decomposition of a matrix. See Chapter 18 for an example.

If A is a matrix that can be put into echelon form *without row interchanges*, then there is a lower triangular matrix L and an upper triangular matrix U such that the following equation holds. $A = LU$.

This code is more complicated than most of the other examples in this appendix. It is designed to be as easy as possible to enter, even at the possible expense of efficiency of calculation. Some of the lines of code such as `next(a,i,j):=` may look like mistakes, but they are not. Begin with a clear screen and **Author** each line *exactly* as it appears below.

```
d_(a):=dimension(a)
i_(n):=identity_matrix(n)
e_(a,i):=element(a,i)
```

```
ee_(a,i,j):=element(a,i,j)
e_(a,k)-ee_(a,k,j)e_(a,i)/ee_(a,i,j)
e_(a,i)/ee_(a,i,j)
if(k=i,#6,#5)
if(k<i,e_(a,k),#7)
down(a,i,j):=vector(#8,k,d_(a))
i>d_(a) or j>d_(a')
sum(|ee_(a,s,j)|,s,i,d_(a))
next(a,i,j):=
h_(a,i,j):=
h_(down(a,i,j),i+1,j+1)
if(ee_(a,i,j)=0,next(a,i,j),#14,#14)
h_(a,i,j):=if(#10,a,#15)
[["Row Interchange Required"]]
next(a,i,j):=if(#11=0,h_(a,i,j+1),#17,#17)
append(a',i_(d_(a)))'
vector(e_(a,k),k,d_(a)-n+1,d_(a))'
l_(a,n):=if(d_(a)=1,a^-1,#20)
upper(a):=h_(a,1,1)
lower(a):=l_(h_(#19,1,1)',d_(a))^-1
```

If A is a matrix that can be put into echelon form *without row interchanges*, then LOWER(A) gives a lower triangular matrix, and UPPER(A) an upper triangular matrix such that LOWER(A)UPPER(A) $= A$. This is the LU decomposition of A.

If a row interchange in encountered, the file ceases calculation and returns the message "Row Interchange Required."

The QR.MTH File

This code produces the QR factorization of A and executes the QR algorithm. See Chapter 18 for an example.

If A is a matrix whose columns are linearly independent, then there is a matrix Q with orthonormal columns and an invertible upper triangular matrix R such that $A = QR$. This is the QR factorization of A. It should be noted that the matrix Q is precisely the result of applying the Gram-Schmidt process to the columns of A.

This code, like the LU code before, is more complicated than most of the other examples in this appendix. It is designed to be as easy as possible to enter. Begin with a clear screen and **Author** each line *exactly* as it appears below.

```
u_(a):=a/|a|
d_(a):=dimension(a)
e_(a,i):=element(a,i)
i_(n):=identity_matrix(n)
(e_(w,k).e_(z,i))e_(z,i)
sum(#5,i,1,k-1)
if(i<k,e_(z,i),u_(e_(w,k)-#6))
vector(#7,i,k)
gram_h(z,w,k):=if(k>d_(w),z,gram_h(#8,w,k+1))
append(a,i_(d_(a')))
q(a):=gram_h([u_(e_(a',1))],a',2)'
r(a):=q(a)'.a
qr(a,n):=iterate(r(z).q(z),z,a,n)
```

If A is a matrix whose columns are linearly independent, then $Q(A)$ returns a matrix with orthonormal columns, and $R(A)$ returns an invertible upper triangular matrix such that $A = Q(A)R(A)$.

The function $QR(A, n)$ executes n iterations of the QR algorithm, returning a matrix that is similar to A and approximately upper triangular. (Be sure to **approX** $QR(A, n)$ rather than **Simplify** it.) See Chapter 18 for an example.